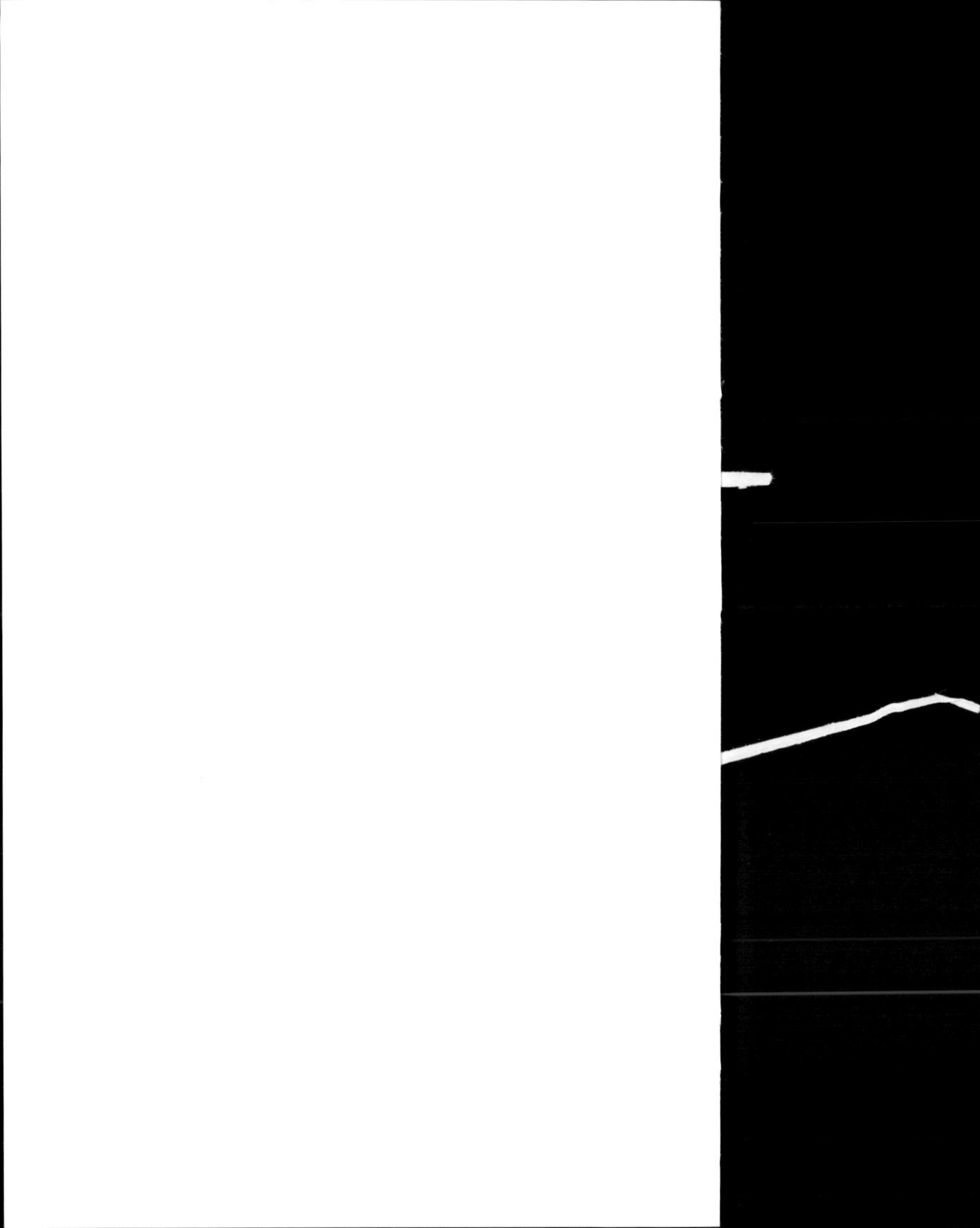

Advance Praise for
McGraw-Hill's Guide to UK Wiring Standards
for Earthing & Bonding

"Although it hardly resembles a wire, the Earth's soil is a conductor, one that you can't ignore. Wherever electricity is delivered, some of it has a good chance of dissipating and residing in the thin upper crust of our planet, especially when the electricity has nowhere else to go. Understanding the relationship between electrical fields and earth is essential to any engineer's success in circuit design and energy delivery. Using concise text and crystal-clear diagrams, David Stockin's book describes exquisitely just what is required to connect wiring to make the circuit you're using work, while the Earth becomes conductor, capacitor, and antenna. It's a subtle business involving soil—dry or moist, rocky or fine—along with metal conductors, complex connectors, and long runs of hollow conduit. If you are responsible for delivering power, or if you just want to know what's going on beneath your feet, read these pages; earth your circuit as he suggests, and you can close the master switch with confidence."

—Bill Nye, "The Science Guy"

————McGraw-Hill's Guide to————

UK WIRING STANDARDS FOR EARTHING & BONDING

David R. Stockin
Manager of Engineering
E&S Grounding Solutions
Hermosa Beach, California

Illustrations by Gilbert Juarez, RevDesign

Contributing Authors:
Michael A. Esparza
Jeffrey D. Drummond
Svetlana Knyazeva-Johnson
Christopher Clemmens
Joseph A. Anderson

Mc
Graw
Hill
Education

New York   Chicago   San Francisco
Athens   London   Madrid
Mexico City   Milan   New Delhi
Singapore   Sydney   Toronto

McGraw-Hill Education books are available at special quantity discounts to use as premiums and sales promotions or for use in corporate training programs. To contact a representative, please visit the Contact Us page at www.mhprofessional.com.

McGraw-Hill's Guide to UK Wiring Standards for Earthing & Bonding

1 2 3 4 5 6 7 8 9 LPI 21 20 19 18 17 16

ISBN 978-1-25-964127-5
MHID 1-25-964127-9

Sponsoring Editor
Michael McCabe

Editorial Supervisor
Stephen M. Smith

Production Supervisor
Lynn M. Messina

Acquisitions Coordinator
Lauren Rogers

Project Manager
Rinki Kaur,
Cenveo® Publisher Services

Copy Editor
Surendra Shivam,
Cenveo Publisher Services

Proofreader
Dipti Barthwal, Cenveo Publisher Services

Indexer
Vikas Makkar, Cenveo Publisher Services

Art Directors, Cover
Jeff Weeks and Gil Juarez

Composition
Cenveo Publisher Services

About the Author

David R. Stockin is Manager of Engineering at E&S Grounding Solutions. A full-time grounding engineer for more than 14 years, he has been the lead electrical grounding/earthing engineer for dozens of data centres used by major search engine companies and government agencies, located from the Americas to Asia and Europe. Mr. Stockin has also been the lead grounding and electrical safety engineer for over a thousand projects involving human safety in high-voltage environments, in both the United States and Canada. He has authored numerous publications on electrical grounding and earthing, including *McGraw-Hill's National Electrical Code® 2014 Grounding & Earthing Handbook*, white papers, and industry proceedings.

About the Contributing Authors

Michael A. Esparza is Director of Sales at E&S Grounding Solutions and has worked in the electrical grounding/earthing industry full time for over 15 years, managing the successful completion of thousands of industry-related projects. He has co-authored numerous publications on electrical grounding and earthing, including *McGraw-Hill's National Electrical Code® 2014 Grounding & Earthing Handbook*, white papers, and industry proceedings.

Jeffrey D. Drummond, P.E., has been practicing engineering for over 18 years and is a licensed Professional Engineer in California, Oregon, Washington, and Idaho. He specializes in industrial power distribution and automation, substations, motor control centres, and industrial control systems, and has been performing grounding studies for these applications since 2002. He is a co-author of *McGraw-Hill's National Electrical Code® 2014 Grounding & Earthing Handbook*. Mr. Drummund has master of engineering and bachelor of science degrees from Harvey Mudd College in Claremont, California, from which he graduated With Distinction. He is a member of Tau Beta Pi, the national engineering honor society.

Svetlana Knyazeva-Johnson has been working full time in the fields of probability statistics, software systems engineering, and cost analysis for nearly 10 years. She holds a bachelor's degree in computer science and engineering from the University of California, Los Angeles, and a master's degree in applied mathematics from the University of Southern California.

Christopher Clemmens has an MBA and a master of accounting degree from the University of Hawaii. He is a tenured diplomat working as a financial management officer for the State Department. Before joining the State Department, Mr. Clemmens was an instructor at the University of Hawaii's Shidler College of Business and at the East-West Center. He is a co-author of *McGraw-Hill's National Electrical Code® 2014 Grounding & Earthing Handbook.*

Joseph A. Anderson has been in the electrical field for over 13 years and is a licensed Master Electrician, certified Electrical Plans Examiner, and certified Commercial Electrical Inspector. He has overseen electrical projects ranging from stray voltage correction/protection to electrical infrastructure improvements and expansions, and he holds additional certifications in corporate safety, team leadership, and fiscal management.

Contents

Acknowledgments

In my first book, *McGraw-Hill's National Electrical Code® 2014 Grounding & Earthing Handbook*, the Acknowledgments section was the most commented-on and praised part. So, I am feeling a bit of pressure to write acknowledgments that are as praiseworthy and full of optimism as the previous ones … but if I'm honest, that is not how I feel.

As before, writing this book has not only been a difficult task, it has also been a great personal learning experience. Most important, it has reaffirmed the value of the people who support me, care about me, provide for me, and generally enable me to be my best. For this book, it turns out that I needed support more than I could have ever imagined.

Writing books is a bit like having children. The first book seems like such a ponderous and yet significantly important task. By the second book, you have your act together and know how to properly prepare and organize for the task at hand. That is exactly how this book was proceeding, in an organized and thoughtful manner … until one day, while writing this book during a retreat at the YMCA of the Rockies, I received a phone call and suffered the single most horrific event in my life … the loss of my son.

It is hard to understand how an event such as the loss of a child can affect your life unless you have lived through it. Today I find myself writing these acknowledgments, just having completed an important milestone, and yet I feel only sadness. I simply do not have the person I want to celebrate this accomplishment with in my life anymore.

This book is dedicated to my son Zachary Robert Stockin. My little boy could walk into a room and his sweet disposition and infectious laughter would put a smile on everyone's face. He was a very important part of a lengthy research project for a revolutionary new class of medicine, which has the potential to save the lives of literally millions of people over the years to come. While he lived a life shorter than most, it was also a life more worthwhile and more fulfilling than most.

I hope you will forgive me for finding my son's life to be worthwhile and extraordinary ... and for expressing the pain of my loss to you. Many of you reading this have no doubt suffered similarly ... perhaps, reading these words will help you find some of the same peace that writing them has brought to me.

That said, I do have some people to whom I need to express gratitude.

To Michael Esparza, who has turned what started out 14 years ago as a shadow of a thought into a career filled with accomplishment, intellectual pursuits, and self-worth, thank you for building something we can all take pride in and hold our heads up high for.

To Gil Juarez, who has always believed in me and supported me: My friend, I know that this was bad timing for you, but your commitment to and extra effort in making this book a reality only makes it more special. Thank you for always being there.

To Jeffrey Drummond, our adventures never seem to stop ... I am proud to call you a partner and I look forward to many more adventures with you in the future. Thank you.

To Svetlana Knyazeva-Johnson, your work ethic, dedication to being the best at what you do, and tireless commitment to our project has been nothing short of inspirational. You may not know this, but numerous times throughout this project, you have motivated me to get it right, to do it better, and to continue to put the effort in no matter how hard it may be. In short, you inspire me, and for that I thank you.

To one of my oldest and dearest friends, Chris Clemmens: Chris, you always seem to know what to say to me, when I need it most. Thank you for your wise counsel and mentorship throughout the years. I really would not know what to do without you.

A big thank you goes to my editor Michael McCabe at McGraw-Hill Education for all the encouragement and guidance over this long project. Thank you for your kind words and for understanding what I was going through during my family's crisis. I look forward to working with you in the future on many more projects.

Thank you, Bill Nye (yes, "The Science Guy"), for all the great conversations, encouragement, and thoughtful science discussions, and for the wonderful opportunities you have given me. Most important, thank you for inspiring me to take a stand and speak my mind about the most important issues. I support you in the goal of thoughtful criticism; through it, we can all help "Change the world!"

I would like to thank Joe Anderson for helping me with this book and telling his story. Joe, your steadfast support and encouragement helped me through the toughest of times. Thank you.

My thanks to Mark Cooper and the management team at the YMCA of the Rockies in Estes Park, Colorado, for providing me with that special writing getaway, and for being there for me during one of the hardest times of my life. No one should ever doubt what amazing people you truly are. I will never forget the kindness and support you showed me during my crisis. Thank you.

To Raul Santiago, thank you for encouraging me, supporting me, and helping me to be my best. I know that when times are at their worst, you will always be there for me. I hope you know that I will be there for you as well, should you need it.

To my good friend, Andrew Son, thank you for keeping my mind straight during the hard times, for keeping me focused on what is most important, and for being the best friend anyone could ever ask for. My life is better because of you.

I ask for the forgiveness of all those who have supported me, consulted with me, and assisted me over the years and whose names I have failed to mention. Thank you.

To my daughter Courtney Stockin, who was always there to support me by picking up the slack, doing what needed to be done without complaint, and taking care of me when I needed it most, I am so proud of the woman you have become. I love you.

As always, to my son Zachary ... you are my heart and my soul.

David R. Stockin

Author's Note to the Reader

As an American who writes technical publications for numerous different countries, I inherently must convert the various terminology used in each country back to what I am most familiar with here in the States, which is of course NFPA 70, the National Electrical Code (NEC).

Undoubtedly, you will notice this tendency to see things from an outside perspective throughout this work. On the downside, this outside perspective may confuse and cause some issues for the reader, so I hope you will accept my apology in advance. On the upside, sometimes an outside perspective is just the thing one needs!

I hope you enjoy this work.

Chapter One

INTRODUCTION TO BS 7671

The British electrical code is one of the oldest and single most well-established electrical codes in the world. At least 27 countries use this important document, word-for-word, for establishing standardized electrical systems within their borders. Numerous other electrical standards from around the world use the British electrical code as the primary guideline for their text. It would be nearly impossible to overstate its importance to the world in helping to provide safe and effective electrical energy to our citizens. The reasons for its success are clear: the British electrical code provides clear and concise guidelines in an organized and logical manner.

The BS 7671 document is broken into numerous parts, chapters, sections, etc. with a document numbering hierarchy as follows:

- Parts—*single digit*
  - Chapters—*two digits*
    - Sections—*three digits*
      - Side-Headings—*three digits plus digits after a decimal point*

Each Part of the BS 7671 code is titled by a single number such as "Part 5." All of the Chapters below Part 5 will be two digits and will start with the same number that they are part of, such as "Chapter 54," which would be the fourth Chapter in Part 5. All of the sections below the chapter will be three digits and will start with the first two numbers from the chapter, such as "Section 542," which would be the second section in Chapter 54. The side-headings simply add a digit after a decimal point, such as "Side-Heading 542.2," which would be the second side-heading in Section 542.

While BS 7671 has 7 Parts, 22 Chapters, plus numerous Sections and Appendices, the bulk of its content can be found in just 4 Parts:

- Assessment of General Characteristics (of electrical systems)
- Protection for Safety
- Selection and Erection of Equipment
- Inspection and Testing

In short, the British electrical code provides guidance for selecting the type of electrical system you have or should have at your facility (Assessment of General

Characteristics), the protection systems and methods you must employ to keep people and equipment safe (Protection for Safety), how to properly select and install the electrical systems you need (Selection and Erection of Equipment), and then how to test the system to ensure its proper operation and safety (Inspection and Testing). This simple and logical format ensures that the British electrical code is an effective and useful document for everyone.

This is very different from the American electrical code, which has been so completely altered by lawyers and courtrooms that one could swear that the point of the document was intentionally designed to confuse everyone who reads it. If this was the case then they have succeeded, because even the writers of the U.S. code themselves cannot seem to get simple engineering terminology straight, and quite often make blatant engineering errors. Fortunately, we do not have to deal with such issues with the British code… well… perhaps there are a few issues, as we will see later in this book.

The national standard in the United Kingdom for low-voltage electrical installations is the British Standard BS 7671 "Requirements for Electrical Installations," which has been in publication since 1882 (in comparison, the U.S. electrical code was first published in 1897). This important document is jointly produced by the Institution of Engineering and Technology (IET) and the British Standards Institution (BSI), and is non-statutory. However, BS 7671 is referenced in several U.K. statutory instruments and is considered to be an official document.

Besides the United Kingdom, the following 27 additional countries are known, or are reliably believed, to also follow BS 7671 as their national electrical code: Botswana, Cyprus, Gambia, Ghana, Gibraltar, Guyana, Kenya, Lesotho, Malawi, Mauritius, Mozambique, Namibia, Nigeria, Rwanda, Seychelles, Sierra Leone, South Africa, Sri Lanka, Saint Vincent and the Grenadines, St. Lucia, Swaziland, Tanzania, Trinidad and Tobago, Uganda, Zambia, and Zimbabwe. The country of Cameroon uses a combination of BS 7671 and the French NFC 15-100. As of 2009, Hong Kong has developed its own national electrical code based largely on BS 7671 called *Code of Practice for the Electricity (Wiring) Regulations 2009*.

Since the 15th edition (1981) of BS 7671, the IET/BSI has been slowly harmonizing the BS 7671 standard to bring it in compliance with the international standard IEC 60364, and is now similar to the current wiring regulations of other European countries.

This has not been the easiest of transitions for the United Kingdom. Most notably in March of 2004, Amendment No. 2 of the 16th Edition of BS 7671 was issued changing the wiring colours to bring it in compliance with IEC standards. This change caused the colour of the neutral conductor to go from black to blue; a confusing issue given that pre-2004 the colour blue was used as L3 in three-phase systems. Black is now L2 in three-phase systems, meaning that electricians must be very cautious when dealing with three-phase wiring in order to stay safe.

The 17th edition of BS 7671, was released in January 2008 and amended in 2011, is notable for requiring residual current devices (RCDs) in Chapter 41 for socket outlets that are for general use by ordinary persons. RCDs greatly improve the level of protection against electrical shock, and keeps the standard in compliance with the latest IEC guidelines.

In January of 2015, the IET/BSI released Amendment No. 3 that introduced a new decimal point numbering system, expanded the use of RCDs, and made a number of other changes throughout the standard typical for an evolving electrical code. This book was written to be in compliance with Amendment No. 3 of the 17th Edition of BS 7671.

In addition to BS 7671, there is the document BS 7430, *Code of Practice for Protective Earthing of Electrical Installations*, that had the latest addition issued in August of 2015 and is titled BS 7430:2011+A1:2015. This book incorporates the 2015 version of BS 7430.

Chapter Two

PART 1: SCOPE, OBJECT, AND FUNDAMENTAL PRINCIPLES

BS 7671 is an excellent electrical code that is broken down into seven parts along with 16 appendices (the actual code itself makes up 282 pages, and the appendices an additional 164 pages, or 37% of the code!). Each 7 of the seven parts is further broken into chapters and sections. However, the bulk of the Code can primarily be found in three sections: Part 4—Protection for Safety, Part 5—Selection and Erection of Equipment, and Part 6—Inspection and Testing. This book will have discussion on those parts, chapters, and sections that relate to earthing conductors and electrodes, equipotential bonding, and the protective conductor (PE), sometimes called the protective "earthing" conductor or PE.

Part 1 of BS 7671 discusses the scope, object, and fundamental principles of the Code. For our purposes, we must understand what is covered under the Code and what is excluded. While BS 7671 provides regulations that are technically non-statutory, in the United Kingdom electrical installations are required to comply with a great number of statutory regulations that reference the regulations found in BS 7671. A near complete list of the applicable laws, legal requirements, and statutory regulations mandating BS 7671 can be found in Appendix 2 of the Code.

In general, BS 7671 applies to electrical installations at residential, commercial, industrial, and public premises that have nominal voltages up to and including 1000 V AC or 1500 V DC. This of course includes other premises and facilities such as medical, agricultural, marinas, portions of caravans, highway systems, mobile equipment, and numerous other areas listed in Chapter 110.1 of the Code.

BS 7671 does *not* apply to electrical systems that distribute electricity to the public, such as those above 1000 V AC. Also excluded from the scope of BS 7671 are railway equipment and facilities, motor vehicles (there are some applicable caravan rules), ships, aircraft, mines, the DC side of cathodic protection systems, and several other areas as shown in Chapter 110.2 of the Code.

Chapter 13 (in Part 1) of BS 7671 talks about the fundamental principles of the Code, and is one of the best parts of the document. For example, the very first principle of BS 7671 is to provide basic protection against direct contact with live parts for of persons and livestock during normal operating conditions. This is done by preventing and/or limiting the current that can pass through the body to a non-hazardous level. The second principle is to provide fault protection against indirect contact due to the failure of basic insulation (abnormal operating conditions). This is done by limiting the magnitude and duration of an unintentional electrical fault, to a non-hazardous level. While numerous other principles are listed in Chapter 13, these first two give you a good idea of how BS 7671 provides general guidelines for the safety of persons and livestock.

Another important principle, as it relates to this book, is found in Part 4 "Protection for Safety" which provides the regulations required for safety of persons. In Chapter 411.3 we are told that the first and best method for protecting persons against electrical shock due to a fault is through protective earthing and protective equipotential bonding. This is a principle that far too few engineers seem to appreciate, thus is our purpose. However, we will talk about that more when we discuss Part 4 of the Code.

Chapter Three

PART 2: DEFINITIONS

Apparent resistivity This is not defined in either BS 7671 or BS 7430. Apparent resistivity is a fictitious soil resistivity assuming theoretically homogenous soil (also referred to as uniform soil or single-layer soil) that is generated from a single measurement (from a Wenner four-point soil resistivity test) to provide an average resistivity from the surface of the earth down to a depth roughly equivalent to the spacing of the probes. To generate an accurate soil model, numerous apparent resistivity measurements should be taken at specific spacing (see Soil modelling).

Arm's reach A zone of accessibility to touch, extending from any point on a surface where persons usually stand or move about to the limits which a person can reach with a hand in any direction without assistance (see Figure 417).

Barrier A part providing a defined degree of protection against contact with live pmts from any usual direction of access.

Basic insulation Insulation applied to live parts to provide basic protection and which does not necessarily include insulation used exclusively for functional purposes.

Basic protection Protection against electric shock under fault-free conditions.
    Note: For low-voltage installations, systems, and equipment, basic protection generally corresponds to protection against direct contact, that is, "contact of persons or livestock with live parts."

Bonding conductor A protective conductor providing equipotential bonding.

Bonding network (BN), {444} A set of interconnected conductive parts that provide a path for currents at frequencies from direct current (DC) to radio frequency (RF) intended to divert, block or impede the passage of electromagnetic energy.

Bonding ring conductor (BRC), {444} A bus-earthing conductor in the form of a closed ring.

Note: Normally, the bonding ring conductor, as part of the bonding network, has multiple connections to the common bonding network (CBN) that improves its performance.

Bypass equipotential bonding conductor, {444}   Bonding conductor connected in parallel with the screens of cables.

Cable ducting   An enclosure of metal or insulating material, other than conduit or cable trunking, intended for the protection of cables which are drawn in after erection of the ducting.

Cable ladder   A cable support consisting of a series of transverse supporting elements rigidly fixed to main longitudinal supporting members.

Cable tray   A cable support consisting of a continuous base with raised edges and no covering. A cable tray may or may not be perforated.

Cable trunking   A closed enclosure normally of rectangular cross-section, of which one side is removable or hinged, used for the protection of cables and for the accommodation of other electrical equipment.

Cable tunnel   A corridor containing supporting structures for cables, joints, and/ or other elements of wiring systems, whose dimensions allow persons to pass freely throughout the entire length.

Cartridge fuse link   A device comprising a fuse element or two or more fuse elements connected in parallel enclosed in a cartridge usually filled with an arc-extinguishing medium and connected to terminations (see Fuse link).

Central power supply system   A system supplying the required emergency power to essential safety equipment.

Central power supply system (low power output)   Central power supply system with a limitation of the power output of the system at 500 W for 3 h or 1500 W for 1 h.
   Note: A low power supply system is normally comprised of a maintenance-free battery and a charging and testing unit.

Circuit   An assembly of electrical equipment supplied from the same origin and protected against overcurrent by the same protective device(s).

Circuit breaker   A device capable of making, carrying, and breaking normal load currents and also making and automatically breaking, under predetermined conditions, abnormal currents such as short-circuit currents. It is usually required to operate infrequently although some types are suitable for frequent operation.

Circuit breaker, linked   A circuit breaker the contacts of which are so arranged as to make or break all poles simultaneously or in a definite sequence.

Circuit protective conductor (cpc)   A protective conductor connecting exposed conductive parts of equipment to the main earthing terminal.

Class I equipment Equipment in which protection against electric shock does not rely on basic insulation only, but which includes means for the connection of exposed conductive-parts to a protective conductor in the fixed wiring of the installation (see BS EN 61140).

Class II equipment Equipment in which protection against electric shock does not rely on basic insulation only, but in which additional safety precautions, such as supplementary insulation, are provided, there being no provision for the connection of exposed metalwork of the equipment to a protective conductor, and no reliance upon precautions to be taken in the fixed wiring of the installation (see BS EN 61140).

Class III equipment Equipment in which protection against electric shock relies on supply at SELV and in which voltages higher than those of SELV are not generated (see BS EN 61140).

Common equipotential bonding system, common bonding network (CBN), {444} Equipotential bonding system providing both protective equipotential bonding and functional equipotential bonding.

Competent person A person who possesses sufficient technical knowledge, relevant practical skills and experience for the nature of the electrical work undertaken and is able at all times to prevent danger and, where appropriate, injury to him/herself and others.

Conducting location with restricted movement A location comprised mainly of metallic or conductive surrounding parts, within which it is likely that a person will come into contact through a substantial portion of their body with the conductive surrounding parts and where the possibility of preventing this contact is limited.

Conductor A material or object that allows electricity or heat to move through it, typically a wire.

Conduit A part of a closed wiring system for cables in electrical installations, allowing them to be drawn in and/or replaced, but not inserted laterally.

Connector The part of a cable coupler or of an appliance coupler which is provided with female contacts and is intended to be attached to the end of the flexible cable remote from the supply.

Consumer unit (may also be known as a consumer control unit or electricity control unit). A particular type of distribution board comprising a type-tested coordinated assembly for the control and distribution of electrical energy, principally in domestic premises, incorporating manual means of double-pole isolation on the incoming circuit(s) and an assembly of one or more fuses, circuit breakers, residual current operated devices or signalling, and other devices proven during the type test of the assembly as suitable for such use.

Continuous operating voltage (Uc) {534} Maximum rms voltage which may be continuously applied to an SPD's mode of protection. This is equal to the rated voltage.

Current-carrying capacity of a conductor   The maximum current which can be carried by a conductor under specified conditions without its steady-state temperature exceeding a specified value.

Current-using equipment   Equipment which converts electrical energy into another form of energy, such as light, heat, or motive power.

Disconnector   A mechanical switching device that, in the open position, complies with the requirements specified for the isolating function.
   Note 1: A disconnector is otherwise known as an isolator.
   Note 2: A disconnector is capable of opening and closing a circuit when a negligible current is either broken or made, or when no significant change in the voltage across the terminals of each pole of the disconnector occurs. It is also capable of carrying currents under normal circuit conditions and carrying for a specified time current under abnormal conditions such as those of short-circuit.

Discrimination   Ability of a protective device to operate in preference to another protective device in series.

Distribution board   An assembly containing switching or protective devices (e.g., fuses, circuit breakers, residual current operated devices) associated with one or more outgoing circuits fed from one or more incoming circuits, together with terminals for the neutral and circuit protective conductors. It may also include signalling and other control devices. Means of isolation may be included in the board or may be provided separately.

Distribution circuit   A circuit supplying a distribution board or switchgear. A distribution circuit may also connect the origin of an installation to an outlying building or separate installation, when it is sometimes called a sub-main.

Distributor   A person who distributes electricity to consumers using electrical lines and equipment that he/she owns or operates.

Double insulation   Insulation comprising both basic insulation and supplementary insulation.

Duct, ducting   (see Cable ducting).

Earth   The conductive mass of the Earth, whose electric potential at any point is conventionally taken as zero.

Earth electrode   Conductive part that may be embedded in the soil or in a specific conductive medium, e.g., concrete or coke, in electrical contact with the Earth.

   Earth plate   A copper plate typically not greater than 1.2 m × 1.2 m (4 ft × 4 ft) connected in parallel vertically and at least 2 m (6 ft) apart with a minimum ground cover of 600 mm (2 ft) of damp soil. Connections to the plate must be of a corrosion-free type, welded or riveted, and covered with a heavy coat of bitumen. See BS 7430-2011+A1-2015 Chapter 9.5.2 for more information.

Earth ring   An earthing electrode, typically composed of 7.42 mm (#2 AWG) or larger bare copper wire at least 6 m (20 ft) in length and buried at least 76 cm (2.5 ft) below grade and in direct contact with the earth. The ground ring circumnavigates a structure or building and is reconnected to itself, typically with multiple connections to the building's or structure's steel rebar and/or the building's steel system, and used in conjunction with standard ground rods. It is considered to be the most effective electrode system in use today.

Earth rod   An earthing electrode composed of stainless steel, copper-coated steel, or zinc-coated (galvanized) steel of at least 1.6 cm (5/8 in.) in diameter with at least 2.4 m (8 ft) of length in direct contact with the earth below the permanent moisture level. This is the most common grounding electrode in use today.

Earth electrode network, {444}   Part of an earthing arrangement comprising only the earth electrodes and their interconnections.

Earth electrode resistance   The resistance of an earth electrode to Earth.

Earth fault current   A current resulting from a fault of negligible imped-ance between a line conductor and an exposed conductive part or a protective conductor.

Earth fault loop impedance (Z$_s$)   The impedance of the earth fault current loop starting and ending at the point of earth fault. This impedance is denoted by the symbol "Z$_s$."
   The earth fault loop comprises the following, starting at the point of fault:
 - The circuit protective conductor
 - The consumer's earthing terminal and earthing conductor
 - For TN systems, the metallic return path
 - For TT and IT systems, the Earth return path
 - The path through the earthed neutral point of the transformer
 - The transformer winding
 - The line conductor from the transformer to the point of fault

Earth leakage current   (see Protective conductor current).

Earth potential (from BS 7430-2011+A1-2015)   Electric potential with respect to the general mass of earth that occurs in, or on the surface of, the ground around an earth electrode when an electric current flows from the electrode to earth (see Potential gradient around earth electrode).

Earth potential rise (from BS 7430-2011+A1-2015)   Voltage between an earth-ing system and reference earth [BS EN 50522:2011] (see Potential gradient around earth electrode).

Earthed concentric wiring   A wiring system in which one or more insulated con-ductors are completely surrounded throughout their length by a conductor, for example a metallic sheath, which acts as a PEN conductor.

Earthed electrical system   An electrical system where the neutral point of the source transformer is connected to earth. TN and TT systems are earthed electrical systems.

Earthing   Connection of the exposed conductive parts of an installation to the main earthing terminal of that installation.

Earthing conductor   A protective conductor connecting the main earthing terminal of an installation to an earth electrode or to other means of earthing.

Electric shock   A dangerous physiological effect resulting from the passing of an electric current through a human body or livestock.

Electrical circuit for safety services   Electrical circuit intended to be used as part of an electrical supply system for safety services.

Electrical equipment   Any item for such purposes as generation, conversion, transmission, distribution, or utilization of electrical energy, such as machines, transformers, apparatus, measuring instruments, protective devices, wiring systems, accessories, appliances, and luminaires.

Electrical installation   An assembly of associated electrical equipment having coordinated characteristics to fulfil specific purposes.

Electrical source for safety services   Electrical source intended to be used as part of an electrical supply system for safety services.

Electrical supply system for safety services   A supply system intended to maintain the operation of essential parts of an electrical installation and equipment:
- For the health and safety of persons and livestock, and
- To avoid damage to the environment and to other equipment.
  Note: The supply system includes the source and the circuit(s) up to the terminals of the electrical equipment.

Electrically independent earth electrodes   Earth electrodes located at such a distance from one another that the maximum current likely to flow through one of them does not significantly affect the potential of the other(s).

Electromagnetic disturbances   BS 7671 does not directly define the term Electromagnetic disturbances. However, it does state the following in 444.1: Electromagnetic disturbances can disturb or damage information technology systems or information technology equipment as well as equipment with electronic components or circuits. Currents due to lightning, switching operations, short-circuits, and other electromagnetic phenomena might cause over-voltages and electromagnetic interference.
  A common definition used in the United States is as follows:

Electrical noise (electromagnetic interference)   Any electromagnetic disturbance that interrupts, obstructs, or otherwise degrades or limits the effective performance of electronics and electrical equipment.

Enclosure A part providing protection of equipment against certain external influences and in any direction providing basic protection.

Equipment (see Electrical equipment).

Equipotential bonding Electrical connection maintaining various exposed conductive parts and extraneous conductive parts at substantially the same potential (see also Protective equipotential bonding).

Exposed conductive part Conductive part of equipment which can be touched and which is not normally live, but which can become live under fault conditions.

Extra-low voltage (see Nominal voltage).

Extraneous-conductive part A conductive part liable to introduce a potential, generally Earth potential, and not forming part of the electrical installation. Examples of extraneous-conductive parts include metal pipes of all types, metal framing, or any other conductive object. See 415.2 and Guidance Note 8, Chapter 6.

Fault A circuit condition in which current flows through an abnormal or unintended path. This may result from an insulation failure or a bridging of insulation. Conventionally the impedance between live conductors or between live conductors and exposed- or extraneous conductive parts at the fault position is considered negligible.

Fault current A current resulting from a fault.

Fault protection Protection against electric shock under single fault conditions.
    Note: For low voltage installations, systems, and equipment, fault protection generally corresponds to protection against indirect contact, mainly with regard to failure of basic insulation. Indirect contact is "contact of persons or livestock with exposed conductive parts which have become live under fault conditions."

Final circuit A circuit connected directly to current-using equipment, or to a socket outlet or socket outlets or other outlet points for the connection of such equipment.

Flexible cable A cable whose structure and materials make it suitable to be flexed while in service.

Functional bonding conductor, {444} Conductor provided for functional equipotential bonding.

Functional earth Earthing of a point or points in a system or in an installation or in equipment, for purposes other than electrical safety, such as for proper functioning of electrical equipment.

Functional extra-low voltage (FELV) An extra-low voltage system in which not all of the protective measures required for SELV or PELV have been applied.

Fuse   A device that, by the melting of one or more of its specially designed and proportioned components, opens the circuit in which it is inserted by breaking the current when this exceeds a given value for a sufficient time. The fuse comprises all the parts that form the complete device.

Gas installation pipe   Any pipe, not being a service pipe (other than any part of a service pipe comprised in a primary meter installation) or a pipe comprised in a gas appliance, for conveying gas for a particular consumer and including any associated valve or other gas fitting.

Hazardous-live-part   A live part which can give, under certain conditions of external influence, an electric shock.

High voltage   (see Voltage, nominal).

Impulse current (Iimp), {534}   A parameter used for the classification test for SPDs; it is defined by three elements, a current peak value, a charge Q and a specific energy W/R.

Impulse withstand voltage, {534}   The highest peak value of impulse voltage of prescribed form and polarity which does not cause breakdown of insulation under specified conditions.

Instructed person   A person adequately advised or supervised by skilled persons to enable him/her to avoid dangers which electricity may create.

Insulation   Suitable non-conductive material enclosing, surrounding or supporting a conductor.

Isolation   A function intended to cut off for reasons of safety the supply from all, or a discrete section, of the installation by separating the installation or section from every source of electrical energy.

Isolator   A mechanical switching device which, in the open position, complies with the requirements specified for the isolating function. An isolator is otherwise known as a disconnector.

Leakage current   Electric current in an unwanted conductive path under normal operating conditions.

Lightning protection zone (LPZ), {534}   Zone where the lightning electromagnetic environment is defined.

Line conductor   A conductor of an AC system for the transmission of electrical energy other than a neutral conductor, a protective conductor or a PEN conductor. The term also means the equivalent conductor of a DC system unless otherwise specified in the Regulations.

Live conductor   (see Live part).

Live part   A conductor or conductive part intended to be energized in normal use, including a neutral conductor but, by convention, not a PEN conductor.

Low voltage   (see Voltage, nominal).

Luminaire   Equipment which distributes, filters, or transforms the light transmitted from one or more lamps and which includes all the parts necessary for supporting, fixing, and protecting the lamps, but not the lamps themselves, and where necessary, circuit auxiliaries together with the means for connecting them to the supply.
   Note: Lamps includes devices such as light emitting diodes.

Main distribution board   (see Distribution board).

Main earthing terminal   The terminal or bar provided for the connection of protective conductors, including protective bonding conductors, and conductors for functional earthing, if any, to the means of earthing.

Main protective bonding conductor (section 544.1)   Used for PME installations with more than one source of supply, the protective conductor provided for protective equipotential bonding. Main protective bonding jumpers are any protective conductor (PE) that is routed from the main earthing terminal (MET) to any other extraneous-conductive-part such as building steel, water pipe, gas line, electronic communication systems, etc.

Maintenance   Combination of all technical and administrative actions, including supervision actions, intended to retain an item in, or restore it to, a state in which it can perform a required function.

Meshed bonding network (MESH-BN), {444}   Bonding network in which all associated equipment frames, racks and cabinets, and usually the DC power return conductor are bonded together as well as at multiple points to the CBN and may have the form of a mesh.
   Note: A MESH-BN improves the performance of a common bonding network.

Midpoint conductor   (see Neutral conductor).

Neutral conductor   A conductor connected to the neutral point of a system and contributing to the transmission of electrical energy. The term also means the equivalent conductor of an IT or DC system unless otherwise specified in the Regulations and also identifies either the mid-wire of a three-wire DC circuit or the earthed conductor of a two-wire earthed DC circuit.

Nominal discharge current (Inspd), {534}   A parameter used for the classification test for Class I SPDs and for preconditioning of an SPD for Class I and Class II tests; it is defined by the crest value of current through an SPD having a current waveform of 8/20.

Nominal voltage   (see Voltage, nominal).

Normally current carrying   A conductor in a circuit that is designed to normally carry electrical currents as part of its typical operation. Line and neutral conductors are normally current carrying.

Normally noncurrent carrying   Conductors or other metallic parts (including exposed metallic parts) that are *not* normally current carrying.

Obstacle   A part preventing unintentional contact with live pm1s but not preventing deliberate contact.

Open-circuit voltage under standard test conditions (Uoc STC)   Voltage under standard test conditions across an unloaded (open) generator or on the DC side of the convertor.

Ordinary person   A person who is neither a skilled person nor an instructed person.

Origin of an installation   The position at which electrical energy is delivered to an electrical installation.

Origin of the temporary electrical installation   Point on the permanent installation or other source of supply from which electrical energy is delivered to the temporary electrical installation.

Overcurrent   A current exceeding the rated value. For conductors the rated value is the current-carrying capacity.

Overcurrent detection   A method of establishing that the value of current in a circuit exceeds a predetermined value for a specified length of time.

Overload current   An overcurrent occurring in a circuit which is electrically sound.

PEL   A conductor combining the functions of both a protective earthing conductor and a line conductor.

PELV (protective extra-low voltage)   An extra-low voltage system which is not electrically separated from Earth, but which otherwise satisfies all the requirements for SELV.

PEN conductor   A conductor combining the functions of both protective conductor and neutral conductor.

Permittivity   Is the measure of the resistance that is encountered when forming an electric field in a medium. In other words, permittivity relates to a material's ability to transmit (or "permit") an electric field.

Phase conductor   (see Line conductor).

Plug   Accessory having pins designed to engage with the contacts of a socket-outlet, and incorporating means for the electrical connection and mechanical retention of a flexible cable.

Potential gradient around earth electrodes   Under fault conditions an earth electrode is raised to a potential with respect to the general mass of Earth that may be calculated from the prospective fault current and the earth resistance of the electrode; this results in the existence of potential differences in the ground around the electrode that might be damaging to telephone and pilot cables, whose cores are substantially at earth potential. Such a risk should be considered mainly in connection with large electrode systems, as at power stations and substations. The potential gradient over the surface of the ground is should also be considered because personnel or livestock can be in contact with two points sufficiently far apart that the potential difference constitutes a danger to life; cattle are most at risk. See BS 7430-2011+A1-2015, Chapter 9.6 for more information.

In the United States, potential gradient around earth electrodes is called a ground potential rise (GPR) or and earth potential rise (EPR).

Ground potential rise (as defined by IEEE Std. 80-2000)   The maximum electrical potential that a (substation) grounding grid may attain, relative to a distant grounding point, assumed to be at potential of remote earth. This voltage, GPR, is equal to the maximum grid current times the grid resistance.

Earth potential rise (as defined in IEEE Std. 367)   The product of ground electrode impedance, referenced to remote earth, and the current that flows through that electrode impedance.

Prospective fault current (Ipf)   The value of overcurrent at a given point in a circuit resulting from a fault of negligible impedance between live conductors having a difference of potential under normal operating conditions, or between a live conductor and an exposed conductive part.

Protective bonding conductor   Protective conductor provided for protective equipotential bonding.

Protective conductor (PE)   A conductor used for some measures of protection against electric shock and intended for connecting together any of the following parts:
 (i) Exposed conductive parts
 (ii) Extraneous conductive parts
(iii) The main earthing terminal
(iv) Earth electrode(s)
 (v) The earthed point of the source, or an artificial neutral

Protective conductor current   Electric current appearing in a protective conductor, such as leakage current or electric current resulting from an insulation fault. See 543.7 for more information.

In the United States, protective conductor current is called objectionable current.

Objectionable current   Electrical current on earthing/grounding and bonding paths that occurs when an improper neutral-to-ground bond (or neutral-to-case) creates a parallel path for neutral current to return to the power supply via metal parts of the electrical system in violation of Art. 250.142. This would be better defined as *objectionable neutral current.*

Protective earthing   Earthing of a point or points in a system or in an installation or in equipment for the purposes of safety.

Protective equipotential bonding   Equipotential bonding for the purposes of safety.

Protective multiple earthing (PME)   An earthing arrangement, found in TN-C-S systems, in which the supply neutral conductor is used to connect the earthing conductor of an installation with Earth, in accordance with the Electricity Safety, Quality and Continuity Regulations 2002.

Protective separation   Separation of one electric circuit from another by means of:
   (i) double insulation, or
  (ii) basic insulation and electrically protective screening (shielding), or
 (iii) reinforced insulation.

Rated current   Value of current used for specification purposes, established for a specified set of operating conditions of a component, device, equipment, or system.

Rated impulse withstand voltage level (Uw), {534}   The level of impulse withstand voltage assigned by the manufacturer to the equipment, or to part of it, characterizing the specified withstand capability of its insulation against over voltages.
   Note: For the purposes of BS 7671 only withstand voltage between live conductors and earth is considered.

Reduced low voltage system   A system in which the nominal line-to-Iine voltage does not exceed I I0 volts and the nominal line to Earth voltage does not exceed 63.5 volts.

Reinforced insulation   Single insulation applied to live parts, which provides a degree of protection against electric shock equivalent to double insulation under the conditions specified in the relevant standard. The term "single insulation" does not imply that the insulation must be one homogeneous piece. It may comprise two or more layers which cannot be tested singly as supplementary or basic insulation.

Reporting   Communicating the results of periodic inspection and testing of an electrical installation to the person ordering the work.

Residual current   Algebraic sum of the currents in the live conductors of a circuit at a point in the electrical installation.

Residual current device (RCD)   A mechanical switching device or association of devices intended to cause the opening of the contacts when the residual current attains a given value under specified conditions.

Residual current operated circuit breaker with integral overcurrent protection (RCBO)   A residual current operated switching device designed to perform the functions of protection against overload and/or short-circuit.

Residual current operated circuit breaker without integral overcurrent protection (RCCB)   A residual current operated switching device not designed to perform the functions of protection against overload and/or short-circuit.

Residual operating current   Residual current which causes the RCD to operate under specified conditions.

Resistance area   (For an earth electrode only.) The surface area of ground (around an earth electrode) on which a significant voltage gradient may exist.

Resistance to ground (RTG)   A DC (0 Hz) measurement of resistance to the flow of electrical current through a grounding electrode in reference to a remote earth ground. In practical terms, the remote earth ground needs to be placed at a distance of at least 10 times the diagonal length of the electrode under test.

Safety service   An electrical system for electrical equipment provided to protect or warn persons in the event of a hazard, or essential to their evacuation from a location.

SELV   (separated extra-low voltage). An extra-low voltage system which is electrically separated from Earth and from other systems in such a way that a single fault cannot give rise to the risk of electric shock.

Shock   (see Electric shock).

Shock current   A current passing through the body of a person or livestock such as to cause electric shock and having characteristics likely to cause dangerous effects.

Short-circuit current   An overcurrent resulting from a fault of negligible impedance between live conductors having a difference in potential under normal operating conditions.

Simple separation   Separation between circuits or between a circuit and Earth by means of basic insulation.

Simultaneously accessible parts   Conductors or conductive parts, which can be touched simultaneously by a person or, in locations specifically, intended for them, by livestock.
   Note: Simultaneously accessible parts may be: live parts, exposed conductive parts, extraneous-conductive-parts, protective conductors, or earth electrodes.

Skilled person   A person with technical knowledge or sufficient experience to enable him/her to avoid dangers which electricity may create.

Socket-outlet   A device, provided with female contacts, which is intended to be installed with the fixed wiring, and intended to receive a plug. A luminaire track system is not regarded as a socket-outlet system.

Soil modelling   Is not defined in either BS 7671 or BS 7430. A calculated comparison of the lateral variations from all of the data points gathered from multiple varied probe spacings (from a four-point soil resistivity test), in order to determine how many soil layers are at a given location, the resistivity of those layers, and at what depth those soil layers reside.

Sphere of influence   Is not defined in either BS 7671 or BS 7430. The hypothetical volume of soil that will experience the majority of the voltage rise of the ground electrode when that electrode discharges current into the soil. The sphere of influence is equal to the diagonal length of the electrode or electrode system.

Spur   A branch from a ring or radial final circuit.

Standby electrical source   Electrical source intended to maintain, for reasons other than safety, the supply to an electrical installation or a part or parts thereof, in case of interruption of the normal supply.

Standby electrical supply system   Supply system intended to maintain, for reasons other than safety, the functioning of an electrical installation or a part or parts thereof, in case of interruption of the normal supply.

Stationary equipment   Electrical equipment which is either fixed or which has a mass exceeding 18 kg and is not provided with a carrying handle.

Step voltage   Is the difference in surface potential experienced by a person bridging a distance of 1 m with the feet, without contacting any grounded object. In practical terms, step voltage is a harmful voltage that can flow between the legs of someone who is walking near a high-voltage source, causing injury to personnel. See IEC 60050-195, 195.05-12.

Supplementary bonding conductors   Are those conductors that are added in addition to the Main Protective Bonding Conductors to ensure fault-protection is maintained throughout a given system.

Surge current, {534}   A transient wave appearing as an overcurrent caused by a lightning electromagnetic impulse.

Surge protective device (SPD), {534}   A device that is intended to limit transient overvoltages and divert surge currents. It contains at least one non-linear component.

Switchboard   An assembly of switchgear with or without instruments, but the term does not apply to groups of local switches in final circuits.

Switchgear   An assembly of main and auxiliary switching equipment for operation, regulation. Protection, or other control of an electrical installation.

System An electrical system consisting of a single source or multiple sources running in parallel of electrical energy and an installation. See Part 3. For certain purposes of the Regulations, types of system are identified as follows depending upon the relationship of the source, and of exposed conductive parts of the installation, to Earth:

- **TN system.** A system having one or more points of the source of energy directly earthed. The exposed conductive-parts of the installation being connected to that point by protective conductors.
- **TN-C system.** A system in which neutral and protective functions are combined in a single conductor throughout the system.
- **TN-S system.** A system having separate neutral and protective conductors throughout the system.
- **TN-C-S system.** A system in which neutral and protective functions are combined in a single conductor in part of the system.
- **TT system.** A system having one point of the source of energy directly earthed, the exposed conductive parts of the installation being connected to earth electrodes electrically independent of the earth electrodes of the source.
- **IT system.** A system having no direct connection between live parts and Earth, the exposed conductive parts of the electrical installation being earthed (see BS 7671 Appendix 9 Figure 9C).
- **Multiple source and DC systems.** see BS 7671 Appendix 9.

Temporary overvoltage (UTOV), {534} A fundamental frequency overvoltage occurring on the network at a given location of relatively long duration.

Note 1: TOYs may be caused by faults inside the LY system (UTQV.LV) or inside the HY system (UTOV.HV)

Note 2: Temporary overvoltages typically lasting up to several seconds, usually originate from switching operations or faults (e.g., sudden load rejection, single-phase faults, etc.) and/or from non-linearity (ferroresonance effects, harmonics, etc.)

Temporary supply unit An enclosure containing equipment for the purpose of taking a temporary electrical supply safely from an item of street furniture.

Testing Implementation of measures to assess an electrical installation by means of which its effectiveness is proved. This includes ascertaining values by means of appropriate measuring instruments, where measured values are not detectable by inspection.

Triplen harmonics The odd multiples of the 3rd harmonic of the fundamental frequency (e.g., 3rd, 9th, 15th, 21st).

Touch voltage Is the potential difference between the GPR of a grounded object and the surface potential, at the point where a person is standing while at the same time having a hand in contact with the grounded object. In practical terms, touch voltage is a harmful voltage that can flow between the hands and feet of someone who is touching a high-voltage source, causing injury to personnel. See IEC 60050-195, 195.05-11, modified.

Verification All measures by means of which compliance of the electrical installation with the relevant requirements of BS 7671 are checked, comprising inspection, testing, and certification.

Voltage, nominal Voltage by which an installation (or part of an installation) is designated. The following ranges of nominal voltage (rms values for AC) are defined:
- **Extra-low.** Not exceeding 50 V AC or 120 V ripple-free DC, whether between conductors or to Earth.
- **Low.** Exceeding extra-low voltage but not exceeding 1000 V AC or 1500 V DC between conductors, or 600 V AC or 900 V DC between conductors and Earth.
- **High.** Normally exceeding low voltage.

Note: The actual voltage of the installation may differ from the nominal value by a quantity within normal tolerances, see Appendix 2.

Voltage, reduced (see Reduced low voltage system).

Voltage band
- **Band I**
  Band I covers:
  - installations where protection against electric shock is provided under certain conditions by the value of voltage;
  - installations where the voltage is limited for operational reasons (e.g., telecommunications, signalling, bell, control, and alarm installations).

  Extra-low voltage (ELV) will normally fall within voltage band I.
- **Band II**
  Band II contains the voltages for supplies to household and most commercial and industrial installations. Low voltage (LV) will normally fall within voltage Band II.

Note: Band II voltages do not exceed 1000 V AC rms or 1500 V DC

Voltage protection level (up), {534} A parameter that characterizes the performance of an SPD in limiting the voltage across its terminals, which is selected from a list of preferred values; this value is greater than the highest value of the measured limiting voltages.

Wiring system An assembly made up of cable or busbars and parts which secure and, if necessary, enclose the cable or busbars.

Chapter Four

PART 3: ASSESSMENT OF GENERAL CHARACTERISTICS

TYPES OF CONDUCTORS

Chapter 31 of Part 3 of BS 7671 deals with the different types of electrical systems, primarily earthed systems and unearthed systems, with additional breakdowns in the earthed category. This is important for us in that each of these systems have different earthing requirements. However, before we get into the different types of electrical systems, we need to spend a minute discussing the different conductors (wires) that will be used in the various electrical system arrangements.

There are two types of conductors, normally current-carrying and normally non-current-carrying. I wish it was that simple; however, it's not quite that easy. The Code has four terms for current-carrying conductors, and seven terms for non-current-carrying conductors. BS 7671 also has four additional terms for conductors that do both.

For the normally current-carrying conductors, we generally think of line conductors and neutral conductors, which is exactly how BS 7671 generally use those terms. However, in certain places the term "phase conductor(s)" is used instead of "line conductor(s)," and the term "midpoint conductor(s)" is used instead of "neutral conductor(s)." As I am sure you know, both line and neutral conductors will carry the full current load of your circuit—one is simply the source of energy and the other the return.

For the normally non-current-carrying conductors, we generally think of the protective conductor (PE); however, BS 7671 has at least five additional terms for the PE used at various points within the Code, plus the term earthing conductor to describe the conductor used to bond below-grade earthing electrodes to the electrical system. These terms are circuit protective conductor (cpc), bonding conductor, bypass equipotential bonding conductor, and protective bonding conductor.

In addition to the terms already discussed above, we have four terms used to describe conductors that both normally carry current and function as a protective conductor:

- PEN conductor
- PEM conductor
- PEL conductor
- Earthed concentric wiring

Obviously as a person who specialises in earthing and human safety, I am opposed to normally current-carrying conductors also being used as protective conductors, but such is the world we live in. The way that PEN/PEM conductors are used, and how the earthing is handled, is the key difference in the various electrical systems. The following chart shows the various terms based on their electrical function.

BS 7671 Conductor Terms

Current-Carrying Conductor		Non-Current-Carrying Conductor	
Terms for a Line Conductor	Terms for a Neutral Conductor	Terms for a Protective Conductor	Terms for Connecting to Earthing Electrodes
Line Conductor (L)	Neutral Conductor (N)	Protective Conductor (PE)	Earthing Conductor
Phase Conductor	Midpoint Conductor (M)	Circuit Protective Conductor (cpc)	
		Bonding Conductor	
		Bypass Equipotential Bonding Conductor*	
		Functional Bonding Conductor	
		Protective Bonding Conductor	
PEL Conductor**	PEN Conductor (PE plus N)		
	PEM Conductor (PE plus M)		
	Earthed Concentric Wiring		

*May conduct some current during normal operations.
**A PEL Conductor is listed as a combination of both a protective conductor (PE) and a line conductor (L), hence the term PEL: However, it is not discussed in BS 7671 or in any IEC document that I could find. It could possibly be a term used for the conductor on a corner grounded delta transformer.

TYPES OF ELECTRICAL SYSTEMS

Arguably, there is nothing more important for the electrical engineer or electrician to know, than what type of electrical system they are dealing with. In general, there are two types of electrical systems, earthed and unearthed, with several variations of the earthed variety. BS 7671 312.2 breaks the various systems down using a simple two-letter code system, with a few subsequent letter modifiers.

The first letter in the code describes the relationship of the electrical system to earth, and is either a "T" for earthed or an "I" for unearthed. The second letter of the code describes the relationship of the exposed conductive parts of the installation to earth, and is either a "T" for a system with an independent earthing electrode bonding the exposed conductive parts to earth, or an "N" for a system where the neutral point of the power system is earthed. This gives us four possible combinations; however, only three are actually possible due to the nature of the laws of physics.

- **TT system** A TT system is an earthed electrical system where the neutral point of the power system (transformer) is earthed at one location, and the exposed conductive parts of the electrical installation are independently earthed at another location, with no connection between the earthing electrodes other than the earth itself. This means that there could be a significant difference in potential between the power system and the electrical installation. TT systems are widely considered unsafe and banned by the electrical codes of many countries. *Circuit breakers don't work in TT systems!*
- **TN system** This is the safest and most common electrical system in the world. A TN system is an electrical system where the neutral point of the power system (transformer) and the exposed conductive parts of the installation are earthed at the same point. This means that there is no difference in potential between the power system and the electrical installation.
- **IT system** An IT system is an electrical system where the neutral point of the power system (transformer) is not connected to earth, or as a significantly high impedance placed between the neutral point and earth in the form of a resistor or inductor. IT systems are most commonly used by distributors in the transmission of electrical energy from transformer to transformer. In industrial settings, IT systems are sometimes used to power three-phase electrical motors. IT systems should not have distributed neutrals.
- **IN system** An IN system is not actually possible because you can't bond the neutral to earth in an unearthed system, can you? If you did, it would become an earthed system and you would get a TN system. It is just physics.

TT systems (312.2.2) are an uncommon type of electrical system, typically found in older installations. In the typical TT system, the high side of the transformer is an IT system (delta), and the low side of the transformer is a TN system (Star of Wye) with the neutral point of the windings bonded to earth via an earthing conductor that is in turn bonded to an earthing electrode. The chassis (exposed conductive part) of the transformer is additionally bonded to the neutral point/ earthing conductor. The neutral is supplied to the consumer's main distribution board and is *not* tied to the consumer's earthing electrode. Figure 4.1 shows the arrangement of a typical TT system.

It should be noted that TT systems are widely considered to be very unsafe (standard circuit breakers may not function) and are actually banned by other electrical codes found around the world, such as the National Electrical Code of the United States. This is because the earth fault loop must pass through the soil of the earth before it gets to the circuit breaker. This is why when testing TT systems, the earth

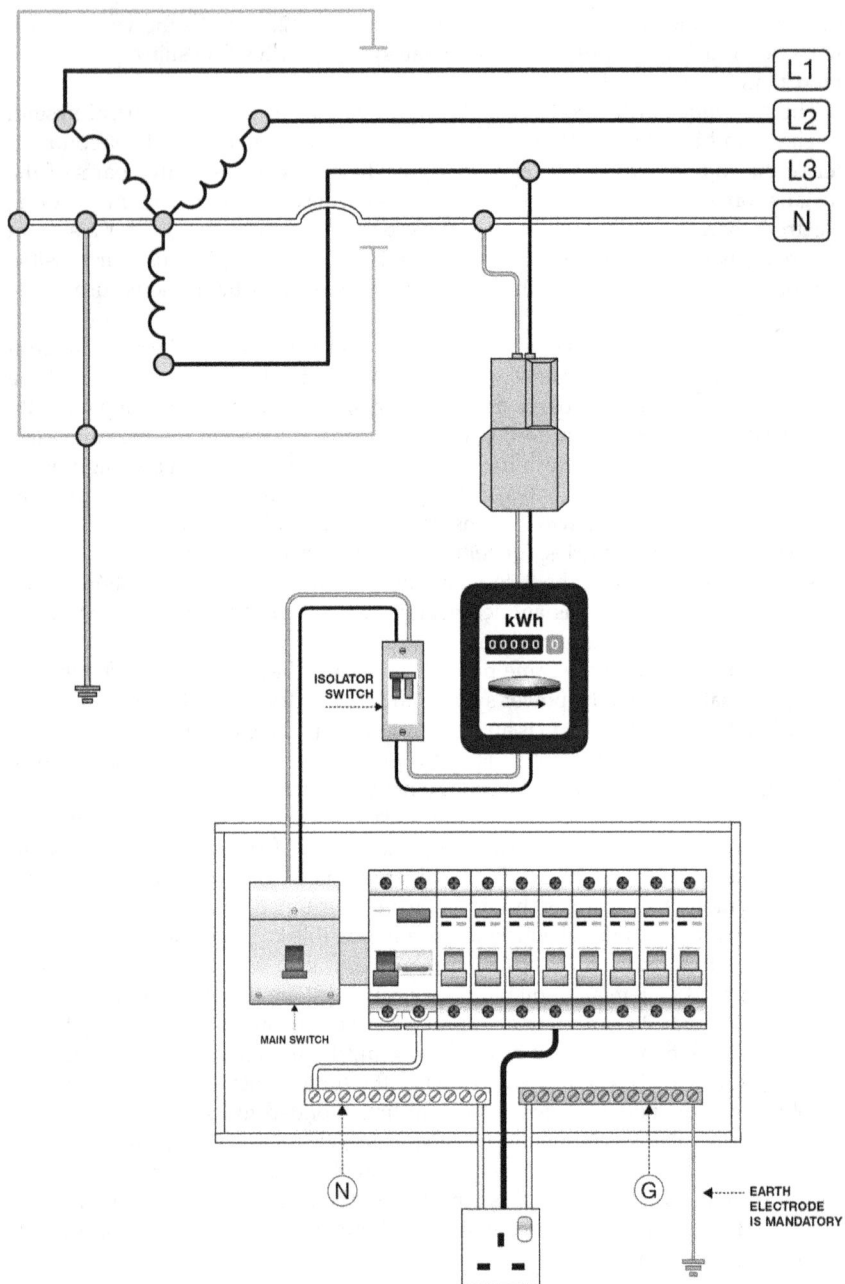

L1
L2
L3
N

ISOLATOR
SWITCH

kWh
00000 0

MAIN SWITCH

N

G

EARTH
ELECTRODE
IS MANDATORY

Figure 4.1 TT system schematic.

fault loop impedance (Z_s) is often grossly over acceptable values. The resistance of the soil/earth can easily be far too resistive (tens to thousands of ohms) for the fault currents to reach a magnitude great enough to trip the circuit breaker and de-energise the circuit. Generally, you will find TT systems in older installations, and rarely in new construction because of the Z_s testing requirements. In Fig. 4.2, you will find a TT system showing the earth fault loop, highlighting how TT systems use the highly resistive soil of the earth as a conductor for the fault current path. *Circuit breakers don't work in TT systems!*

Today, with the advent of new technologies such as RCDs, residual current monitoring, and insulation fault monitoring, TT systems can now be made safe for persons and livestock. Currently, BS 7671 has not mandated the use of such high-tech safety devices for TT systems, but if you do have an older installation, these devices are a great way to make the system safe for persons. However, even with such advanced safety and monitoring devices, there will still be a significant difference in potential between the power system and the electrical system, which is not only bad for the safety of persons, but bad for the telecommunication systems,

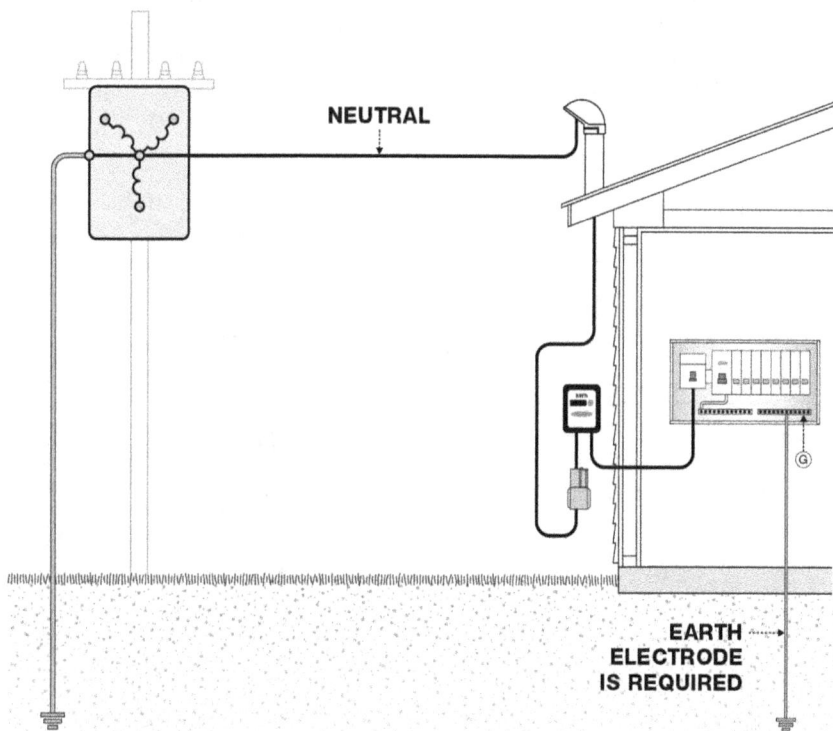

Figure 4.2 TT system fed from utility pole. Note phase conductors have been removed from this drawing for clarity.

television systems, data lines, and so on, that may be connected to your building. Another solution for many TT systems is a simple bonding jumper installed at the main distribution board and you can convert a TT into a TN system as quick as you like.

In Fig. 4.3, we see a TT system wired through a below-grade connection, while in Fig. 4.2, we see a TT system wired using an above-grade connection.

Of note, there has been a big push lately by the manufacturers of these protection devices, which insist that TT systems are superior to TN systems and are the wave of the future. I for one do not buy it, as it seems these claims are motivated less by science, and more by future profit.

IT systems (BS 7671 Appendix 9) are most commonly found in use by the electrical energy distributor to transmit power from power plant to substation, from substation to substation, and from substation to the last transformer before the consumer. IT systems are also used in industrial settings with large three-phase motors where a single transformer is feeding a single motor, or where power is being distributed from transformer to transformer. IT systems typically use delta formation transformers that have no common neutral point for each winding (and should not have any distributed neutrals). Therefore, the electrical system is not connected to earth and is considered "unearthed." The chassis (exposed conductive part) of the delta transformer is bonded to an earthing conductor, which is in turn bonded to an earthing electrode, for the safety of persons.

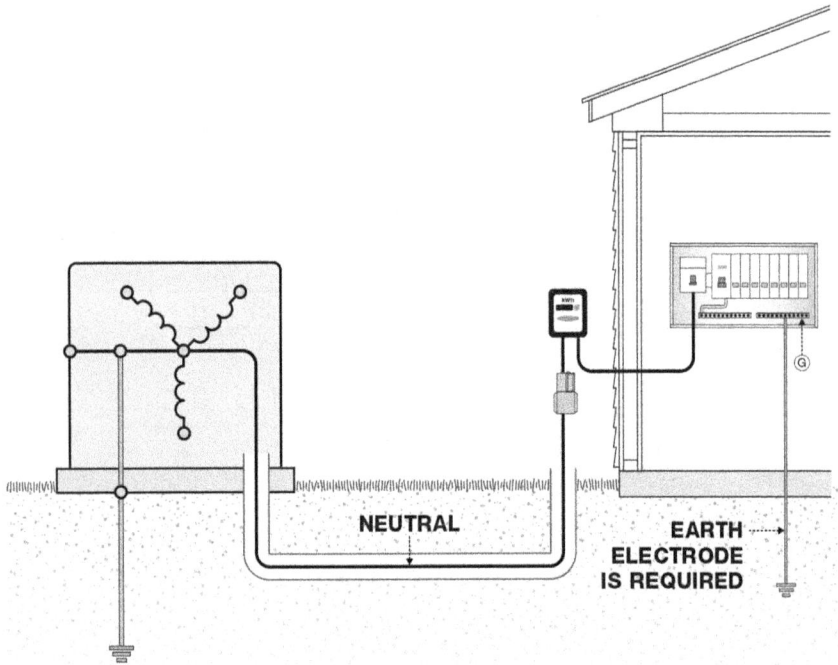

NEUTRAL

EARTH
ELECTRODE
IS REQUIRED

Figure 4.3 TT system fed from a pad-mounted transformer. Note phase conductors have been removed from this drawing for clarity.

IT systems suffer from a similar issue as TT systems, in that they use the highly resistive soil of the earth as a conductor for the fault current path. This is why IT systems are generally only used by distributors and industrial settings, with a series of advanced monitoring and detection mechanisms that are designed to detect such faults and automatically disconnect power.

Note: It is possible to tap the midpoint of one of the winding in a delta transformer and develop a neutral for distribution. However, this generates severally off-balance loads with all the safety problems of a TT system, and is rarely used anywhere. In BS 7671 Part 4, Chapter 41, Regulation 411.6.1, we have a strongly worded warning against distributing neutrals in an IT system.

TN systems (312.2.1) are by far the most common type of electrical system in use around the world at the consumer level. In the typical TN system, the high side of the transformer is an IT system (delta), and the low side of the transformer is a TN system (Star of Wye) with the neutral point of the windings bonded to earth via an earthing conductor that is in turn bonded to an earthing electrode. The chassis (exposed conductive part) of the transformer is additionally bonded to the neutral point/earthing conductor. The neutral is supplied to the consumer's main distribution board and is tied to the consumers earthing electrode.

TN systems can use one of two ways (actually three ways as we will soon see) to distribute the neutral, either using a combined PE and neutral in the form of a single PEN conductor ("C"), or using two separate conductors in the form of a single PE and a single neutral conductor ("S").

The subsequent letter(s), if any, in the code describe the arrangement of the neutral and protective conductors. The letter "C" indicates that the neutral and protective conductors are combined into a single PEN conductor (hence the "C" for combined). The letter "S" indicates that the neutral and protective conductors are separate (hence the "S"). It is possible to have an electrical system where the source of energy uses a PEN conductor ("C") to provide power to the installation, where the PEN conductor is separated into individual neutral and protective conductors ("S"). This gives us four additional TN electrical systems:

- **TN-C**   A TN-C system is a TN system where the neutral and protective earth conductors are combined into a single PEN conductor, and are kept as a PEN conductor throughout the system (from transformer to final circuit). A TN-C system installation is very unusual to find, as it would *not* be safe for use in wiring a home or business. The TN-C system is only for use in a directly connected single-load installations such as a water pump, light pole, and so on.

- **TN-S**   A TN-S system is a TN system where the neutral and protective earth conductors are kept separate throughout the system (from transformer to final circuit).

- **TN-C-S (PME)**   This is the most common form of electrical system in use today. A TN-C-S (PME) is a TN system that uses a PEN conductor ("C") from the power system (transformer) provided by the distributor, and then splits the PEN conductor into separate neutral and protective earth conductors ("S") at the main distribution board. So you would have "C" from transformer to distribution board, and "S" from distribution board to final circuit. PME stands for protective multiple earthing, which means that the PEN conductor on the distributor side of the system is earthed at multiple points. You can think of a single distributor transformer supplying multiple homes as an example of a TN-C-S (PME) system.

- **TN-C-S (PNB)**   A TN-C-S (PNB) is a TN system that uses a PEN conductor ("C") from the power system (transformer) provided by the distributor, and then splits the PEN conductor into separate neutral and protective earth conductors ("S") at the main distribution board. So you would have "C" from transformer to distribution board, and "S" from distribution board to final circuit. PNB stands for protective neutral bonding and is used when the distributor's transformer feeds a single consumer. PNB simply means that the TN-C-S system is earthed at a single point (versus multiple points as in a PME) usually close to the source.

There are currently no subsequent letters in use for TT and IT systems.

TN-C SYSTEMS

A TN-C system is a TN system where the neutral and protective earth conductors are combined into a single PEN conductor, and are kept as a PEN conductor throughout the system (from transformer to final circuit). In other words, there are only current-carrying conductors used throughout the circuit.

TN-C systems are highly unusual electrical systems and are in fact not defined under BS 7671 Part 3, Assessment of General Characteristics. In other words, you should *not* wire your home or business using a TN-C system. This is because use of a TN-C system throughout a home would place hazardous return currents (neutral currents) on exposed conductive-metal parts throughout the structure.

A TN-C system should probably never be used at all at a system-wide level (at the level of loads it is used all the time, such as laptops, hand-held tools, etc.); however, it could be made safe for a directly connected single-load installation, such as a water well pump, an isolated light pole, or other such isolated system where hazardous electrical currents will not have the option of flowing through exposed non-current-carrying conductive-metal parts as a return path to the source. One can imagine a transformer with a disconnect switch feeding a submerged water pump, or possibly street lights hanging off of wood poles, as possible TN-C applications.

The TN-C system is briefly discussed in BS 7430 and *Guidance Note 8.*

In a TN-C system, the earth fault loop does not rely on the soil of the earth as a conductor of the fault energy. The entire electrical fault will stay on conductive metal conductors for its entire path (in this case the PEN conductor), providing a low-impedance route and maximizing the fault current enabling the circuit breaker to function and disconnect the power.

TN-S SYSTEMS

The second option for a TN system is to use two separate conductors for the protective and neutral conductors. Figure 4.4 shows the arrangement of a typical TN-S system.

As you can see in the figure, in a TN-S system the earth fault loop does not rely on the soil of the earth as a conductor of the fault energy. The entire electrical fault will stay on conductive metal conductors for its entire path (in this case the PE),

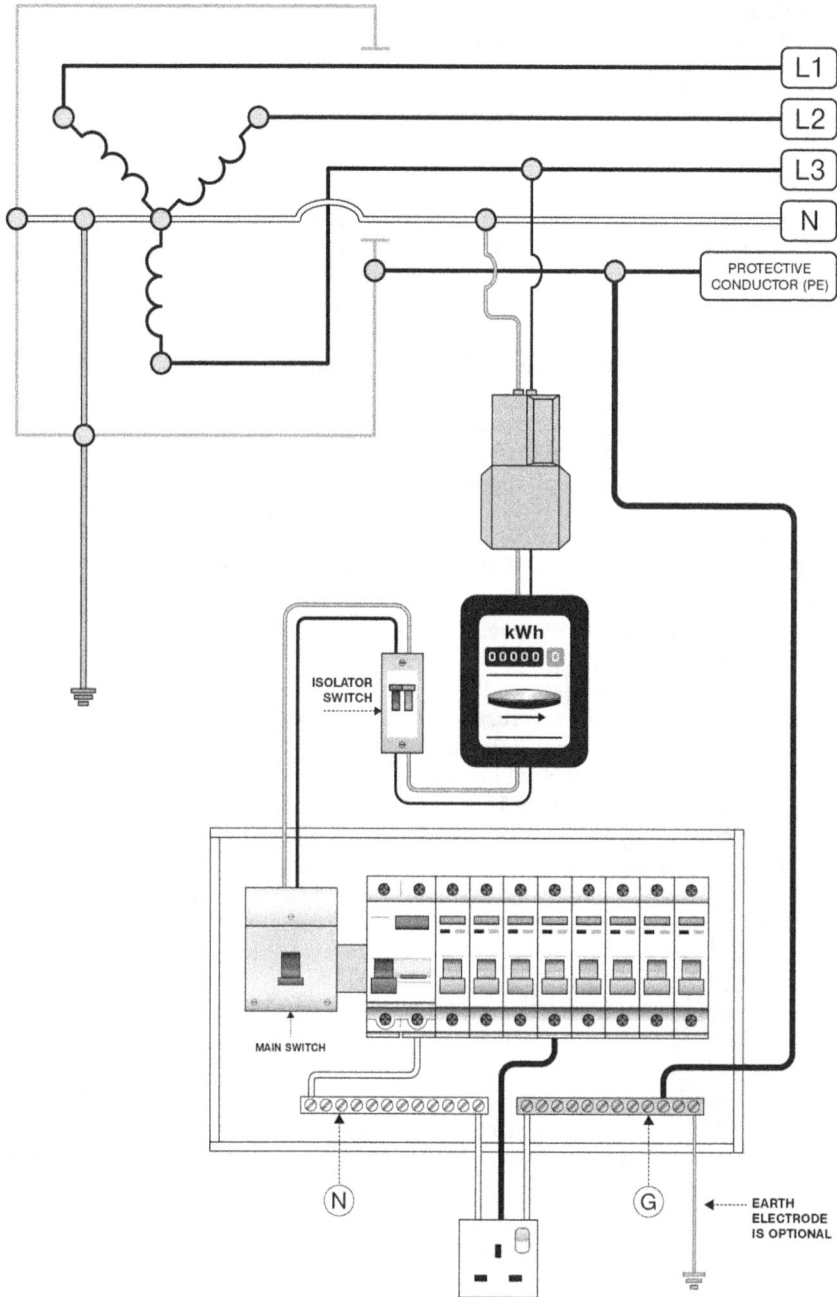

Figure 4.4 TN-S system schematic.

providing a low-impedance route and maximizing the fault current enabling the circuit breaker to function and disconnect the power.

TN-C-S SYSTEMS

The final and third option for a TN system is to a combination of a TN-C and a TN-S, called a TN-C-S system, which is the most commonly used system in the world. In a TN-C-S, the distributor provides a TN-C up to the consumer's main distribution board. Then in the consumer's distribution board, the electrical system is changed to a TN-S system. The combination of the two different systems in one electrical system makes the TN-C-S system.

But we are not quite done yet with the possible combinations. There are actually two versions of a TN-C-S system, and it all depends on how many consumer installations (main distribution boards) are being serviced by a single distributor power system (transformer). When the distributor's power system is supplying multiple installations, you will in essence be sharing the same neutral (PEN conductor) with the other consumers. Each consumer's main distribution board will have its own earthing electrode(s), meaning that this type of TN-C-S system will have multiple earthing, or PME. We call this arrangement a TN-C-S (PME).

Figure 4.5 shows the arrangement of a typical TN-C-S (PME) system.

As you can see in the figure, in a TN-C-S (PME) system the earth fault loop does not rely on the soil of the earth as a conductor of the fault energy. The entire electrical fault will stay on conductive metal conductors for its entire path (in this case the PEN conductor), providing a low-impedance route and maximizing the fault current enabling the circuit breaker to function and disconnect the power.

In the second version of a TN-C-S system, when the distributor's power system is supplying a single installation, the same neutral (PEN conductor) is *not* being shared with other consumers, and therefore will have only its own earthing electrode(s), meaning that this type of TN-C-S system will have a single earthing, or protective neutral bonding (PNB). We call this arrangement a TN-C-S (PNB).

Figure 4.6 shows the arrangement of a typical TN-C-S (PNB) system.

As you can see in Fig. 4.7, in a TN-C-S (PME) system the earth fault loop does not rely on the soil of the earth as a conductor of the fault energy. The entire electrical fault will stay on conductive metal conductors for its entire path (in this case the PEN conductor), providing a low-impedance route and maximizing the fault current enabling the circuit breaker to function and disconnect the power.

The chart on p. 35 will help you to see all of the various electrical systems listed in BS 7671 and in *Guidance Note 8: Earthing & Bonding*. The corresponding figure numbers from BS 7671 and *Guidance Note 8* have been listed for your reference.

In conclusion, for the vast majority of cases you will be dealing with a TN-C-S system where the distributor will provide a PEN conductor to the distribution board, where the PEN conductor will be split in two to provide separate neutral and protective conductors throughout the building. If you happen to run into a TT system, you will almost certainly not be able to pass the earth fault loop (Z_s) testing requirements and you should consider converting the system to

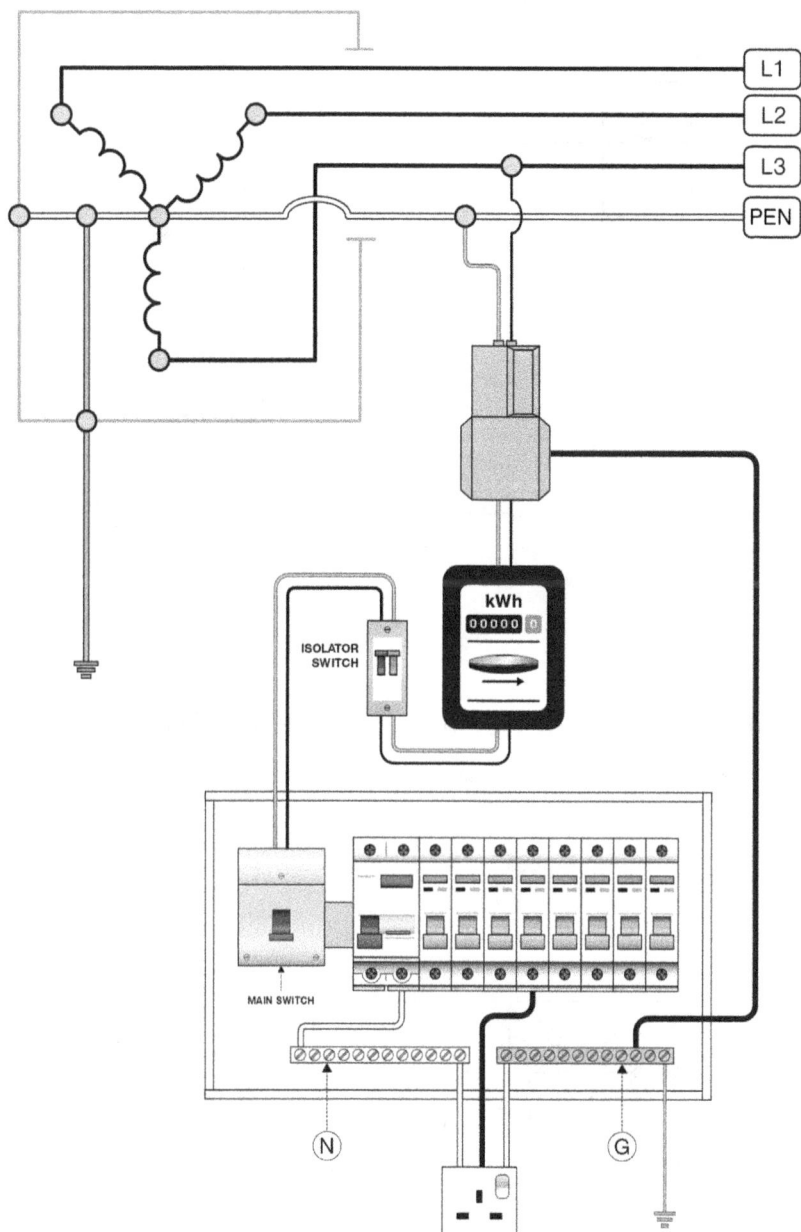

Figure 4.5 TN-C-S (PME) system schematic.

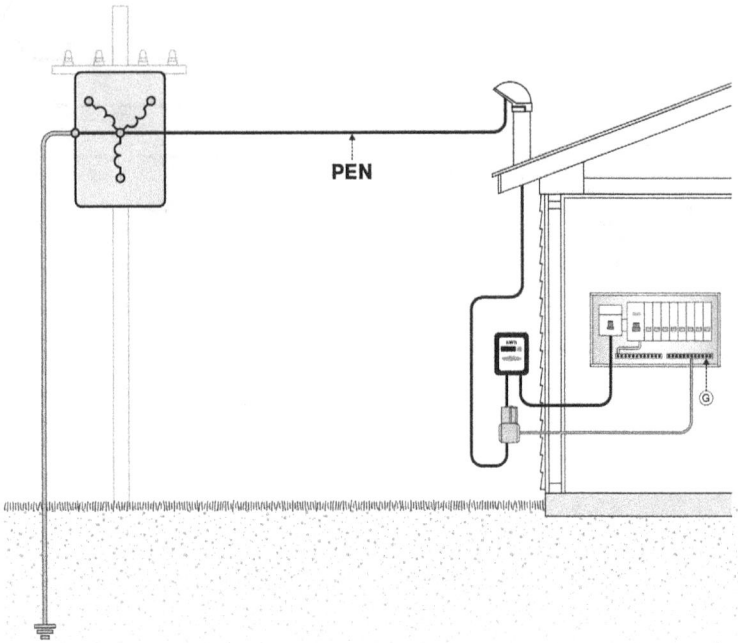

Figure 4.6 TN-C-S (PME) fed from utility pole. Note phase conductors have been removed from this drawing for clarity.

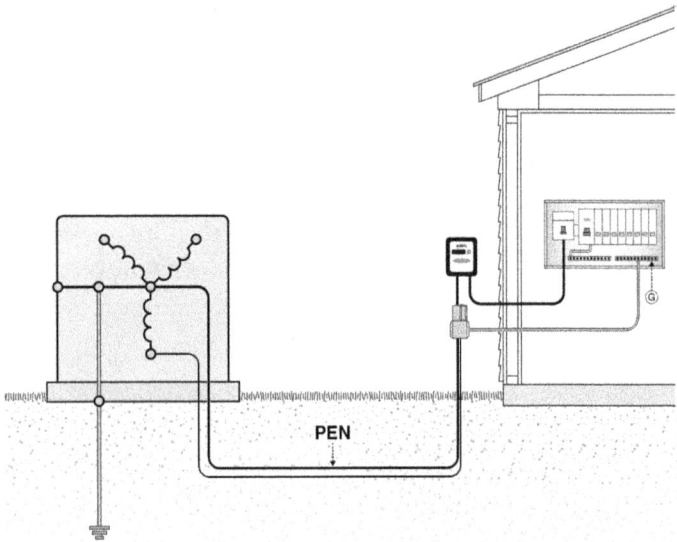

Figure 4.7 TN-C-S (PME) fed from pad-mounted transformer. Note phase conductors have been removed from this drawing for clarity.

Types of Electrical Systems in BS 7671

(T) Earthed				(I) Unearthed	
Exposed Conductive Parts Relationship to Earth				Exposed Conductive Parts Relationship to Earth	
(T) Connected Directly to Earth		(N) Connected to the Earthed Neutral		(T) Connected Directly to Earth	
Arrangement of Neutral and Protective Conductors				Arrangement of Neutral and Protective Conductors	
(S) Neutral and Protective Conductors are Separate	(C) Neutral and Protective Conductors area Single Conductor (PEN Conductor)	(S) Neutral and Protective Conductors are Separate	(C) Neutral and Protective Conductors are a Single Conductor (PEN Conductor)	(S) Neutral and Protective Conductors are Separate	(C) Neutral and Protective Conductors are a Single Conductor (PEN Conductor)
TT		TN		IT	
Figure 3.10		Figure 9B		Figure 9C	
		TN-S	TN-C		
		Figure 3.8	Fig. 4.1 Guid. Note 8		
		TN-C-S (PME)			
		Figure 4.9 Guidance Note 8			
TT (DC System)		TN-C-S (DC System)		IT (DC System)	
Figures 9J and 9K		Figures 9H and 9I		Figures 9L and 9M	
		TN-S (DC System)	TN-C (DC System)		
		Figures 9D and 9E	Figures 9F and 9G		

a TN system, or installing RCDs and other advanced monitoring systems for safety of persons.

Each of these electrical systems has a variety of differing requirements for safety devices, such as circuit breakers, fuses, RCDs, and so on, which are covered under BS 7671 Part 4.

All of these systems, including IT systems, have earthing electrode requirements for the exposed conductive parts that must be met, as required in BS 7671 Part 5, Chapter 54.

Chapter Five

PART 4: PROTECTION FOR SAFETY

Part 4 of BS 7671 deals with the protection and safety of persons and livestock. It is broken down into four chapters and one annex, each placing various requirements to protect persons from: electric shock, thermal effects, overcurrent, and voltage and electromagnetic disturbances. For our purposes, the majority of the earthing and bonding requirements are found in Chapters 41 and 44, protection from electric shock and protection from voltage and electromagnetic disturbances. However, we do have a few things to discuss in regards to the other chapters.

PROTECTION AGAINST ELECTRIC SHOCK, OVERCURRENT, VOLTAGE DISTURBANCES, AND ELECTROMAGNETIC DISTURBANCES

This chapter places the requirements for circuit breakers, fuses, residual current devices (RCD), earthing of exposed conductive parts, earthing of transformers, and more. Each of the various types of electrical systems has its own requirements that must be fulfilled in order for the system to be safe. We will take a minute to discuss each of the safety mechanisms that are required.

The basic fundamentals for electric safety are quite simple:

1. Make hazardous-live-parts inaccessible to persons and livestock
2. Earth all exposed conductive parts and bond them together so that there are no differences in potential and to protect against electromagnetic disturbances
3. Provide protection devices to automatically disconnect power in the event of a fault
4. Provide devices to automatically remove overvoltages

The best way to protect persons and livestock from electric shock is to earth all exposed conductive parts and bond them together to ensure that there are no differences in potential.

BS 7671 recognizes that the various types of electrical systems require different types of protection mechanisms in order to make them safe for persons and livestock. However, every type of electrical system must have the following protections:

- Basic protection—Chapter 41
- Fault protection—Chapters 41 and 44
  - Protective earthing
  - Protective equipotential bonding
- Automatic disconnection of supply—Chapters 41 and 43
  - Circuit breaker or fuse
  - Residual current device (RCD)
- Surge protection device (SPD)—Chapter 44

The fundamental rule of protection against electric shock is to keep hazardous-live-parts inaccessible. This is what the term *basic protection* means. An electrical system with basic protection is one that has physical barriers, insulation, obstacles, and other mechanisms that prevent contact from live parts. There are of course some rules for electrical systems that are located in areas with restricted access to only skilled persons, to remove basic protections. However, other than those conditions, all electrical systems must employ basic protection at all times.

The second requirement for all electrical systems is to employ fault protection. BS 7671 tells us that the best way to protect persons and livestock from electric shock is to earth all exposed conductive parts and bond them together to ensure that there are no differences in potential. This means that the metal chassis of electrical enclosures, electrical equipment, metal conduit, switchboards, switch gear, and so on must be bonded together, and then connected to an earthing electrode. The protective conductor (PE) is the conductor used to make these mandatory bonding connections.

As discussed in Part 3, there are a number of terms BS 7671 uses to describe the conductors used for the protective earthing and protective equipotential bonding requirements:

- Protective conductor (PE)
- Circuit protective conductor (cpc)
- Bonding conductor
- Bypass equipotential bonding conductor
- Functional bonding conductor
- Protective bonding conductor

In addition, an "Earthing Conductor" is used to connect the earthing terminal in the main distribution panel to the earthing electrode or earthing-electrode system. Nevertheless, the earthing conductor is technically not a conductor used for bonding and we will discuss this more in Part 5. However, the requirement found in 411.3.1.1 for protective earthing is telling us that all exposed conductive parts must be earthed, and the earthing conductor is how we will do it.

Each of the conductors used for protective earthing (411.3.1.1) and equipotential bonding (411.3.1.2) must comply with the requirements found in Chapter 54, such as material types, cross-sectional area, and so on.

In 411.3.1.1, we are told that in order to meet the requirements of protective earthing, we must connect all exposed conductive parts to a protective conductor, which is in turn connected back, in practice, to the earthing terminal in the distribution board. This is also called the main earthing terminal or "MET" for short. The earthing terminal must then be connected to the earthing conductor/earthing electrode, and must also have:

- **TN Systems** (411.4.2)—A protective conductor or PEN conductor bonded back to the neutral or midpoint of the power system (transformer)
- **TT Systems** (411.5.1)—No additional connections required
- **IT Systems** (411.6.2)—No additional connections required

In 411.3.1.2, we are told that in order to meet the requirements of protective equipotential bonding, we must connect all extraneous-conductive parts to a protective conductor, which is in turn connected back, in practice, to the earthing terminal in the distribution board. The extraneous-conductive parts include the following:

- Water pipes
- Gas pipes
- Other pipework and ducting
- Heating and air conditioning systems
- Exposed metallic structural parts of the building
- Lightning protection systems (LPS) in accordance with BS EN 62305
- Additional buildings if they are served by the same electrical installation
- Metallic sheaths of telecommunication cables (permission of cable owner required)
- Underground structural metalwork embedded in foundations, such as steel rebar in concrete (415.2.1)

In Part 5 Regulation 542.2.3, we are allowed to use the underground structural metalwork embedded in foundations (concrete-encased steel rebar) as an earthing electrode. However, 415.2.1 requires the connecting the steel rebar in concrete to the protective equipotential system as well, because many concrete foundations have vapour barriers installed between the concrete and the earth (soil) to prevent water seepage. When this occurs, the vapour barrier will electrically insulate the underground structural metalwork embedded in foundations from the earth/soil, meaning that it is no longer a valid earthing electrode. However, the steel rebar in the concrete foundation is very capable of generating a hazardous potentials (voltages) to persons and livestock, hence why 415.2.1 and 411.3.2 require you to connect the underground structural metalwork embedded in foundations to the main earthing terminal (MET) of your electrical system.

While the Code does not specifically mention any other required components to be bonded, Part 2 of BS 7671 does define an extraneous-conductive part as: "A conductive part liable to introduce a potential, generally earth potential, and not forming part of the electrical installation." In Guidance Note 8: Earthing & Bonding Section 6.1, we find two different formulas for calculating whether an extraneous-conductive part is capable of transferring a hazardous voltage (potential) to a person or livestock. We also find similar instructions in BS IEC 60050-195 and IEV 195-05-11 modified and –C12. If you find yourself needing to calculate transferred potentials, often called step and touch voltages, due to a voltage hazard to persons at your site, you should really consult a professional to ensure your site is safe.

A good rule of thumb is to stretch both arms out (2 m) and see if you can touch any conductive object with one hand and any part of the electrical system, including electrical apparatus that may be plugged into an outlet, with the other. If you can, than you should probably consider that conductive object an "extraneous-conductive part," and bond it to the protective equipotential system.

Regardless of your ability to touch a conductive object or not, the following is a list of items that you should bond to the MET of your electrical system (your protective equipotential system):

- Fire sprinklers (411.2.1.2)
- Cable television systems (CATV)
- Broadband systems
- Alarm systems
- Optical fibre systems
- Metal fences (typically mandatory in high-voltage areas)
- Street lights and other lighting poles

It should be noted that in the *On-Site Guide* for BS 7671:2008 (2011), Figure 2.4.3(ii) specifically shows the bonding of nearby metal fences and streetlights when using power from a generator where it was not possible to install an earthing electrode.

As stated earlier (this can just not be stated enough), the best way to protect persons and livestock from electric shock, is to earth all exposed conductive parts and bond them together to ensure that there are no differences in potential; BS 7671 Regulation 411.3.1.1 and 411.3.1.2 require this for all electrical systems.

The third part of protection against electric shock is to provide devices that will automatically disconnect electrical power in case of a fault. There are three different protection devices that are discussed in the Code that can be used for automatic disconnection, two of those devices are required (primary) for all circuits, and the third is an additional form of protection that is added above and beyond the first:

- Required (primary)
  - Fuse, or
  - Circuit breaker
- Additionally Required (Secondary)
  - Residual current device (RCD)

An RCBO (residual current operated circuit breaker with overcurrent protection) is both an RCD and a circuit breaker in a single device, and qualifies for both the primary and secondary circuit protection.

We seldom see fuses in use at a distribution board these days. It is more likely to find a fuse in an electronics cabinet, electronic device, or in a ring circuit BS 1363 socket/plug. So for all practical purposes, the requirements in Chapter 41 for protection against electric shock, and in Chapter 43 the protection against overcurrent, are both met by installing a properly sized circuit breaker in the distribution board for each circuit in the electrical system.

RCDs are considered an additional form of protection and must be used in conjunction with a circuit breaker or fuse (411.4.4 and 415.1.2). The current edition of BS 7671 requires that all socket outlets of 20 amps or less, that are intended for general use (i.e., most residential installations) have RCD protection, with only a few exemptions. Additionally, cables buried in walls and outdoor mobile equipment with a current rating of 32 amps or less must be protected by an RCD.

There are a few important rules one should know about RCDs based on the specific type of electrical system you have:

- **TN-C systems**—Per 411.4.4, you may not use RCDs in a TN-C system. This is because a TN-C system combines the functions of the neutral and protective conductors into a single PEN conductor, so the RCD cannot properly detect current differentials.
- **TN-C-S systems**—Per 411.4.4, a PEN conductor may not be used on the load side of an RCD. Additionally, the circuit's protective conductor must be installed on the load side of the RCD.
- **TT systems**—Per 411.5.2, in a TT system, the RCD becomes primary and the circuit breaker secondary. This is due to the resistive nature of the earth fault loop (Zs) inherent in TT systems. *Circuit breakers don't work in TT systems!*
- **IT systems**—Per 411.6.3, RCDs are just one of five types of devices that can be employed to provide circuit protection.

There are of course numerous other rules and regulations for the installation of circuit breakers and RCDs to be found in BS 7671, but that is not the point of this book now, is it? This is just a simple summary to help clarify the requirements of Part 4. Our concern is on the earthing and bonding of electrical systems, which is what we will be getting into in the next chapter.

It is important to know that are some electrical systems and electrical devices that have been double insulated, have reinforced insulation, or have an extra-low voltage that under certain circumstances do not require a protective conductor. The most common item that is double insulated would be your typical extra-low voltage direct current (DC) transformer you use to charge your cell phone. There are a good number of rules regarding extra-low voltage systems, double or reinforced insulation, and electrical separation for such devices and electrical systems to be found in BS 7671 Regulations 411.7, 412, 413, and 414.

While not a complete list, the following chart summarizes the various electrical safety methods required in BS 7671:

Required Protecive Measures		General Requirements	TN Systems	TN-C System	TN-C-S System	TT Systems	IT Systems	FELV System	SELV System	PELV System	Reduced Low Voltage System
Basic Protection		410, 410.3.2, 411.1, 411.2						411.7.2	414.4.1	414.4.1	411.8.2
Double Insulation or Reinforced Insulation			Required under certain conditions. See Section 412 for more information.						414.4.1	414.4.1	
Electrical Separation			Required under certain conditions. See Section 412 for more information.						414.1.1, 414.4.1	414.1.1, 414.4.1	
Fault Protection	Protective Earthing	411.1, 411.3.1.1						411.7.3			
	Protective Equipotenital Bonding	411.1, 411.3.1.2									
Automatic Disconnection of Supply	Circuit Breaker or Fuse*	411.3.2.1, 431.1.1, 434.2	411.3.2.3, 411.4.4	411.3.2.3, 411.4.4	411.3.2.3, 411.4.4	411.3.2.4, 411.5.2*	411.6.3				411.8.3*
	RCD*	411.3.3	411.4.4	NOT allowed 411.4.4	411.4.4*	411.5.2*	411.6.3				411.8.3*
	Secondary Fault Disconnection					Not Required but Highly Encouraged	411.6.4				
Monitoring Devices	Insulation Monitoring (IMD)						411.6.3.1				
	Residual Current Monitoring (RCM)						411.6.3.2				
	Insulation Fault Locations System						411.6.3.2				

Surge Protection Device (SPD)	443.1.1								
Power Supply Earthing		411.4.2	411.4.2	411.4.2	411.5.1	411.6.1		414.4.1	411.8.4.2
Earthing of Exposed Conductive Parts		411.4.2	411.4.2	411.4.2		411.6.2	NOT earthed per 414.4.4	414.4.1	
Protective Conductor (PE) Requirements for Circuits						411.7.5			411.8.5

*Notes
An RCBO is both a circuit breaker and an RCD and will fullfill both requirements.
An RCD is required for general use sockets of 20A or less, and circuits used for outdoor mobile equipment of 32A or less.
Basic Protection—Live parts are physically not accessible. Conductive parts that are accessible shall not be hazardous under normal or fault conditions.
Protectice Earthing—bonding of exposed conductive parts.
Protective Equipotential Bonding—bonding of "extraneous" conductive parts—water pipes, gas pipes, etc.
For TN-C-S Systems that use an RCD, the PEN conductor shall NOT be used on the load side. The PE must be bonded to the PEN on the source side. See 411.4.4
For TT Systems only one (1) form of protective device is required, see 411.5.2.
For Reduced Low Voltage Systems only one (1) form of protective device is required, see 411.8.3.
An RCD may NOT be the sole form of protection for a circuit per 415.1.2.

The last section of Part 4, protection for safety, is the Annex A444 (Informative) Measures against electromagnetic disturbances. This informative annex tells us that for most dwellings and small commercial buildings where limited electronic equipment is used, specifically electronic equipment that is not interconnected by signal cables, that the protective conductors routed in the power circuit may provide adequate protection from electromagnetic disturbances. However, for commercial and industrial buildings containing multiple interconnected electronic applications, a common equipotential bonding system is useful in order to comply with the electromagnetic compatibility (EMC) requirements of the various electronic systems.

Annex A444 gives us four types of equipotential bonding networks:

- In section A444.1.1 we find the protective conductors (PEs) in a star network. This is the recommended model for dwellings and small commercial buildings with limited electronics. The star network is simply the PE routed with the power circuit out to a typical plug.

  Note: In this scenario, there is only the PE network installed, there is no additional equipotential bonding system.

- In section A444.1.2, we find the multiple meshed bonding star network. In this network, we have multiple small equipotential bonding meshes installed in the structure and connected together using the star network from A444.1.1. The star network can be made of the protective conductors (PE) in the power circuit, or can be an additional *functional bonding conductor*. You could imagine a small laboratory office building with an equipotential mesh in the server room bonding all the computer servers together, and then a separate equipotential bonding mesh in the laboratory room bonding all the electronic measurement equipment together, and then using a star network to connect the individual meshes back to the MET for the building.

  Note: In this scenario, there are two earthing systems: the PE network, and an additional equipotential bonding system.

- In section A444.1.3, we find a common meshed bonding star network. This network is similar to the network found in section A444.1.2, except that as the title indicates, there is simply a single common meshed network, plus the star protective conductors (PE) network routed in the power circuit from A444.1.1. This type of system provides the highest form of protection from electromagnetic disturbances and is often installed at data centres where lightning strikes are a concern.

  Note: In this scenario, there are two earthing systems: the PE network, and an additional equipotential bonding system.

- The last scenario found in section A444.2 is an equipotential bonding network in buildings with several floors. This section tells us that whatever type of equipotential bonding network you might have on a given floor [bonding ring conductor (BRC), common meshed bonding network, multiple star/mesh bonding network, or star network], you must have at least two Chapter-54-approved conductors bonding the floors together, plus two approved conductors bonding the equipotential bonding networks to the earthing electrode system for the building. In A444.3, we find a recommendation that common meshed bonding star networks on each floor are the best option.

Note: In this scenario, there are two earthing systems: the PE network, and an additional equipotential bonding system.

If you are involved in a building with an equipotential bonding system designed to protect sensitive electronic equipment, it is recommended that you review BS EN 50310 "The application of equipotential bonding and earthing in buildings with information technology equipment."

Chapter Six

PART 5: SELECTION AND ERECTION OF EQUIPMENT

Part 5 of BS 7671 is the largest core part of the Code, containing 80 pages of material (Part 7, Special Installations or Locations, is 91 pages, but it is made up of numerous individual issues and does not cover a core concept of the Code). This important section literally tells us how to choose the correct equipment for our electrical system and how that gear is to be properly installed. A very important part indeed! Part 5 is broken down into six chapters:

    Chapter 51, Common Rules, which has six sections

    Chapter 52, Selection and Erection of Wiring Systems, which has ten sections

    Chapter 53, Protection, Isolation, Switching, Control, and Monitoring, which has nine sections

    Chapter 54, Earthing Arrangements and Protective Conductors, which has four sections

    Chapter 55, Other Equipment, which has seven sections

    Chapter 56, Safety Services, which has eleven side headings

Obviously, for the purposes of our project here, Chapter 54 will be key to understanding earthing as it relates to BS 7671, *Guidance Note 8,* and BS 7430. However, there are a few details in the other chapters in Part 5 related to earthing and the protective conductor (PE) that we need to know about.

CHAPTER 51, COMMON RULES

There is not a lot in the Common Rules chapter that concerns earthing and grounding. As you can imagine, this is where you find all the requirements mandating that your electrical equipment is compliant with the various codes and regulations, meets voltage, current, frequency, and power requirements, is compatible with other electrical equipment, is capable of withstanding impulse voltages and other external influences, and so on.

What interests us in relation to earthing and protective conductors is the requirements found in 514.3 and 514.4, which are the sections that govern the identification of conductors. In 514.3, we see a requirement to identify the various conductors and in 514.4 we are told that the conductors are to be identified with specific colours. As you will recall, in March 2004, BS 7671 updated the colour coding of conductors to comply with International Electrotechnical Commission (IEC) standards. This has presented its fair share of problems, especially since the neutral conductor now has the same colour (blue) as what used to be phase-3 (L3), and what used to be the neutral (black) is now phase-2 (L2). You can imagine that an electrician must be extra cautious when dealing with blue and black conductors as they could be hot or they could be earthed! Yikes!

The following chart from Section 514.4, Identification of Conductors by Colour, shows all of the various colour codes.

Alphanumeric	New Colour Codes (post-2004)	Old Color Codes (pre-2004)
L1	Brown	Red
L2	Black	Yellow
L3	Grey	Blue
Neutral	Blue	Black
Protective conductor (PE)	Green and Yellow	Green and Yellow
Functionial earthing conductor (equipotential bonding)	Cream	—
L+ (DC Circuit)	Brown	—
L– (DC Circuit)	Grey	—
M "Mid-Wire" (DC Circuit)	Blue	—
Extra Low Voltage (ELV) Line Circuits/Conductors	Brown, Black, Red, Orange, Yellow, Violet, Grey, White, Pink or Turquoise	—
ELV Neutral (N) or Mid-Wire (M)	Blue	—
Earthed Protective Extra Low Voltage (PELV)	Blue	—

For earthing and bonding, the neutral or midpoint conductor is to be blue, and the protective conductor is to be green with a yellow stripe. Functional earthing conductors, those used to provide a second earthing connection in the form of an equipotential bonding network, are to be cream coloured. In 514.4.6, we see that bare conductors can also be used; however, when necessary they are to be identified with tape, sleeve, disc, or paint.

In 514.4.5, we find that the single colour green is forbidden from use under BS 7671.

We also find under 514.6.1 that certain conductors are excluded from the colour-marking requirements, specifically: the concentric conductor on a cable, metal sheaths, or armour on cables (when used as a PE), and several types of PE conductors used for bonding exposed or extraneous conductive parts.

There are also some required warning notices for earthing and bonding. In 514.13.1, we find that certain earthing and bonding conductors are required to be permanently affixed with a visible durable label (see BS 951 for labelling requirements) that states:

"Safety Electrical Connection—Do Not Remove"

This label must be applied at the following locations:

1. On every earthing conductor where it connects to an earth electrode.
2. On every bonding conductor where it connects to an extraneous conductive part, such as a metal water tank, radio tower, or other object as required under 415.2.
3. On every earthing and bonding connections to a main earthing terminal (MET) that is outside and/or separate of the main switchgear, such as a copper earthing bar.

There are also some additional warning notices required for those electrical systems that are earth-free, that is, *not* earthed, such as an IT (delta) system (see 312.2 and 312.4). When dealing with such an IT earth-free system, a prominent and durable warning notice shall be installed that says the following:

The protective bonding conductors associated with the electrical installation in this location MUST NOT BE CONNECTED TO EARTH.

Equipment having exposed conductive parts connected to earth must not be brought into this location.

There are of course numerous other warning notices to be found under 514.10 to 514.15, but these other warning notices generally fall outside of the earthing and bonding preVue. However, if you are dealing with equipment with conductor colours from pre-2004, you are required to install a warning notice stating the following:

CAUTION

This installation has wiring colours to two versions of BS 7671. Great care should be taken before undertaking extension, alteration, or repair that all conductors are correctly identified.

While BS 7671 does not specifically require it, in the spirit of the Code one can assume that equipment from the United States or another country that has non-standard conductor colours (in the United States, line wires are typically L1-black, L2-red, and L3-blue, the neutral is white, and the earthing conductor is solid green) would also require a warning notice.

CHAPTER 52, SELECTION AND ERECTION OF WIRING SYSTEMS

This chapter deals primarily with ensuring that conductors are properly rated, routed, and protected for the electrical systems and environments where they will be installed. While this chapter has few direct requirements for earthing and bonding conductors, it does govern numerous requirements for all conductors (including earthing and bonding conductors) in an electrical system.

To start with, we find in 521.5.1 that the line conductors, neutral conductors, and protective conductors from a given circuit are required to be installed and routed in the same enclosure (raceway) with each other. This is for numerous reasons, but primarily for the reduction of the electromagnetic effects of the line conductor. We also find in 521 numerous requirements for proper installation of flexible cables, securing of cables, routing of multiple circuits in enclosures and conduit, prevention of electromechanical and mechanical stresses, and so on. In short, the protective conductor must be included in the circuit wiring and must meet all of the same installation requirements as line and neutral conductors.

In 522, we find a series of specific installation requirements for circuits that may have special needs in relation to extreme ambient temperatures, external heat sources, presence of water or high humidity, risk of impact from solid foreign bodies/objects, risk from corrosive and/or polluting substances, impacts, vibration, flora and mold growth, fauna, solar radiation, seismic action, extreme weather (air movement), etc. In all cases, every cable and conductor including the earthing and bonding conductors, must meet the requirements found in this section.

Starting in Section 523, we find the area where the Code describes the proper sizing (cross-sectional area) of conductors to ensure their ability to handle the likely current loads; or the current-carrying capacity of a conductor/cable. In short, normal-current-carrying conductors (line and neutral) must be sized so that the temperature of the insulation does not exceed a given limit, typically 70°C, but can be as high as 105°C given certain insulation sheathing. We also find an excellent account of how to determine the proper sizing of a neutral conductor for polyphase loads, single-phase loads, and where harmonics may be present. The rule of thumb is of course that the neutral and line conductors should be of the same size. If you are sizing these conductors, you should thoroughly review BS 7671 Section 523 and BS 7769.

In Side Heading 523.6.4, we find two rules that directly apply to earthing and bonding. The first is that PEN conductors [where the neutral (N) and protective conductor (PE) are combined into a single conductor] are to be sized as if they were a neutral conductor per Section 523.

The other rule in 523.6.4 has a bit of unfortunate wording as the Code states "Conductors which serve the purpose of protective conductors only are not to be taken into consideration." I say this is unfortunate, as it would be very easy to construe this statement as meaning that are no requirements for the proper sizing of protective conductors. What the Code is trying to say is that you may not include the cross-sectional area of the protective conductor in the calculations for determining the size of the neutral (or PEN) conductor. This is because under normal operating conditions the protective conductor is non-current-carrying. During normal operating conditions, the entirety of the load currents must be carried by the line and neutral conductors (and occasionally the PEN conductor). The protective conductor is a back-up safety conductor designed to balance voltages and handle the current load for a very short period of time; just long enough to trip your circuit breaker or blow the fuse.

Side Heading 523.6.4 of the Code would be better worded if it said something like, "The protective conductor is a non-current-carrying conductor under normal operating conditions, and therefore may not be considered part of the circuit as those conductors carrying load current. Refer to Section 543 for determining the proper cross-sectional area of conductors which serve the purpose of protective conductors only."

Chapter 52 closes out with instructions regarding proper electrical connections (joints) of conductors, including earthing and bonding conductors. Issues such as ensuring that there is no mechanical strain on conductors, that they are properly protected, and that they are properly sealed, and so on can be found in Sections 526 to 529. For our purposes, it should be noted that all conductor connections are required under 526.3 to be assessable for inspection, testing, maintenance, and cleaning (Section 529), with a few exceptions: joints designed to be directly buried (such as those used in earthing systems), welded joints, soldered joints, brazed joints, joints made via an appropriate compression tool, joints made by a manufacturer not intended to be inspected or maintained, those joints complying with BS 5733 and marked as maintenance-free by a symbol with the letters "MF" surrounded by a circle, and several other exceptions.

CHAPTER 53, PROTECTION, ISOLATION, SWITCHING, CONTROL, AND MONITORING

This chapter of BS 7671 deals primarily with selection and installation of residual current devices (RCDs), fuses, surge protection devices (SPDs), and other monitoring systems. There are only a few earthing and bonding requirements that concern us in this chapter.

In regard to RCDs, we find in 531.2.5 that any RCD installed in a circuit that normally has a protective conductor, must have that PE installed onto the RCD, regardless of the current levels on the circuit (even when operating at current levels below 30 mA). This side heading further tells us that an RCD without the proper protective conductor is considered *not* sufficient to provide the required circuit protection.

Obviously, the protective conductor installed on the RCD provides standard circuit protection as it would for any other circuit (i.e., the proper operation of a circuit breaker during electrical fault conditions). However, it has not always been widely accepted that an RCD requires a PE to enable the RCD to function properly. This is because an RCD without a PE will in fact function as it should in most cases. However, RCDs without a protective conductor are more likely to suffer from accidental disconnects, slower or delayed disconnects, or even failure to disconnect given certain conditions. Most importantly, an RCD without a protective conductor, could suffer from an electrical fault (short) placing hazardous voltage on the enclosure, and without the PE, it might not trip the circuit breaker.

Bottom line, an RCD is *not* a replacement for your protective conductor network; it is a supplement to the PE network. You are required to install the protective conductor on your RCD for installations on earthed electrical circuits.

One of the other areas covered in Chapter 53 is SPDs. We find in 534.2.2 the requirement to connect the output of the SPD directly to the MET or to a protective conductor, whichever is closer. There are numerous scenarios presented in the Code, along with numerous illustrations, all of which require a connection to the facilities MET, either directly or through an existing protective conductor. The cross-sectional area (sizing) of all conductors related to SPDs can be found in 534.2.10.

There are three scenarios that use copper (or equivalent) cross-sectional areas:

- If the line conductors of the circuit to be protected have a cross-sectional area equal to or greater than 4 mm^2, then the conductors for the SPD must have a cross-sectional area of not less than 4 mm^2.
- If the line conductors of the circuit to be protected have a cross-sectional area less than 4 mm^2, then the conductors for the SPD must have a cross-sectional area equal to the line conductor.
- If the SPD is a Type 1 SPD for use in a structural lightning protection system, then the minimum cross-sectional area will be 16 mm^2.

We also find in 534.2.9 certain requirements to ensure that the earthing connecting from the SPD to the MET is as short a routing as possible, and without excessive bends or loops (which could generate a magnetic cross-couplings thereby increasing impedance).

The last issue in Chapter 53 related to earthing and bonding can be found in Section 538.2, which relates to the fault monitoring for IT electrical systems (delta power). In an IT electrical system, only the exposed conductive parts of the system are earthed. The windings in the electrical transformer are not earthed. With IT electrical systems, 538.2 and 538.3 require the installation of an insulation monitoring device (IMD) (along with electrical fault detecting/location devices, RCDs, and more in accordance with 411.6) in accordance with BS EN 61557-9. For our purposes here, we need to know only that in 538.3 the IMD shall be connected between earth and a live conductor.

CHAPTER 54, EARTHING ARRANGEMENTS AND PROTECTIVE CONDUCTORS

In Chapter 54, we finally get to the heart of the enterprise of this book, which is the connection to earth. There are three reference codes that all provide guidance in the area of earthing: BS 7671, *Guidance Note 8*, and BS 7430 (there is of course the *On-Site Guide*, but that is just an extension of the Code). In the following pages, you will find an abbreviated analysis combining these three regulations, while emphasizing BS 7671, with the hopes that this will give you a more complete picture of the total earthing requirements of your electrical system.

All three codes have different purposes. BS 7671 details the "where and how" of earthing and is fairly straightforward by dictating which electrical systems require a connection to earth and what earthing electrodes are acceptable. It tells us the required sizing of earthing conductors, required protections, and where these earthing conductors need to be connected.

In *Guidance Note 8: Earthing & Bonding*, we get additional details and illustrations regarding the required connections to the various electrical systems, and further instruction on how to properly size earthing conductors. We see additional illustrations and details regarding the mandatory connections to various metallic objects such as building structural steel, water pipes, lightning protection systems, data communication systems, and so on. We also see how to earth and bond multiple buildings together and details regarding the making of proper connections and labelling.

In BS 7430, *Code of Practice for Protective Earthing of Electrical Installations*, we see similar details from the other codes regarding the earthing of the various electrical systems; however, we gain new insights on how to properly connect and earth electrical generators to our electrical systems. For the most part, BS 7430 introduces us to how to measure soil resistivity, how to measure the resistance of earth electrodes, and lists equations for calculating (estimating) the resistance of an electrode in a given soil resistivity.

As you can see, these three codes when looked at as a whole give us a very good picture of our earthing requirements. The most common point of all three of these codes is that the earthing arrangement on our electrical systems is vital. Before working on any electrical system, we must know if our system is a TN-S, a TN-C-S, an IT, or a TT system. Each of these electrical systems has a different earthing arrangement as discussed in Part 3 of the Code (see relevant chapter in this book).

TN-S SYSTEMS

In a TN-S system (arguably the safest electrical system), the earthing electrode is connected to the star or midpoint of the power transformers winding, and two earthed conductors are provided to the main switchgear, a neutral and a protective conductor. The protective conductor is supplied directly from the transformer and must have a cross-sectional area of at least 16 mm^2.

A typical TN-S system is found when the incoming utility feed has three conductors, a hot and neutral conductor, surrounded by an earthed sheath (cable sheath earth) acting as a separate PE conductor. It is also possible to have the separate PE conductor in the form of a regular wire-type conductor instead of a sheath, or even to have a buried conductor in the form of a ground ring acting as the PE conductor. In a TN-S system, the only connection between the neutral and the protective earth (PE) conductors occurs at the midpoint of the transformer and at no other point in the system.

The fault return path (fault-current loop) in a TN-S system is through the protective earth conductor to the transformer.

Earthing electrodes are not required to be installed at the consumer's main switchgear for TN-S and TN-C-S systems. However, there is no reason not to install these earthing electrodes, and there are in fact many reasons you should install them. Take a quick look through BS 7671 *Guidance Note 8* and you will find numerous illustrations showing such "means of earthing" to the MET. Additionally, certain countries may require you to install earthing electrodes at the consumer's switchgear.

The National Electrical Code (NEC) and numerous international standards actually require a minimum of two earthing electrodes at the consumer's main switchgear for TN-S and TN-C-S systems.

TN-S ILLUSTRATIONS

Figure 6.1 shows a typical TN-S system from the transformer to the final circuit (socket-outlet). TN-S systems are arguably the safest electrical system available. However, TN-S systems are uncommon, as most electrical distributors do not want to bear the financial costs of providing the additional protective conductor to the consumer's installation.

The important part that makes this a TN-S system, is that the distributor (electric utility provider) is providing three conductors to the installation: line wire, neutral wire, and a protective conductor wire.

Note that there is no neutral-to-ground bond at the distributor's cut-out.

The earth rod electrode installed at the consumer's earth bar (marked with a circle G) is optional for TN-S systems, but highly recommended.

Figure 6.2 shows the same TN-S system illustrated in Fig. 6.1, but in a layout format. The line conductor has been removed from the illustration for clarity reasons, and only shows the neutral and protective conductor (protective earth or PE).

In this case, the distributor (electric utility provider) is providing overhead power to the installation, via three conductors to the installation: line wire (not shown for clarity), neutral wire, and a protective conductor wire that is *not* in contact with the earth.

Note that there is no neutral-to-ground bond at the distributor's cut-out.

An earth rod electrode (not shown) installed at the consumer's earth bar (marked with a circle G) is optional for TN-S systems, but highly recommended.

Figure 6.3 shows the same TN-S system illustrated in Fig. 6.1, but in a layout format. The line conductor has been removed from the illustration for clarity reasons, and only shows the neutral and protective conductor (protective earth).

In this case, the distributor (electric utility provider) is providing underground power to the installation, via three conductors to the installation: line wire (not shown for clarity), neutral wire, and a protective conductor wire that is *not* in contact with the earth.

Note that there is no neutral-to-ground bond at the distributor's cut-out.

An earth rod electrode (not shown) installed at the consumer's earth bar (marked with a circle G) is optional for TN-S systems, but highly recommended.

Figure 6.4 shows the same TN-S system illustrated in Fig. 6.1, but in a layout format. The line conductor has been removed from the illustration for clarity reasons, and only shows the neutral and protective conductor (protective earth).

In this case, the distributor (electric utility provider) is providing overhead power to the installation, via three conductors to the installation: line wire (not shown for clarity), neutral wire, and a protective conductor wire that is buried and in direct contact with the earth.

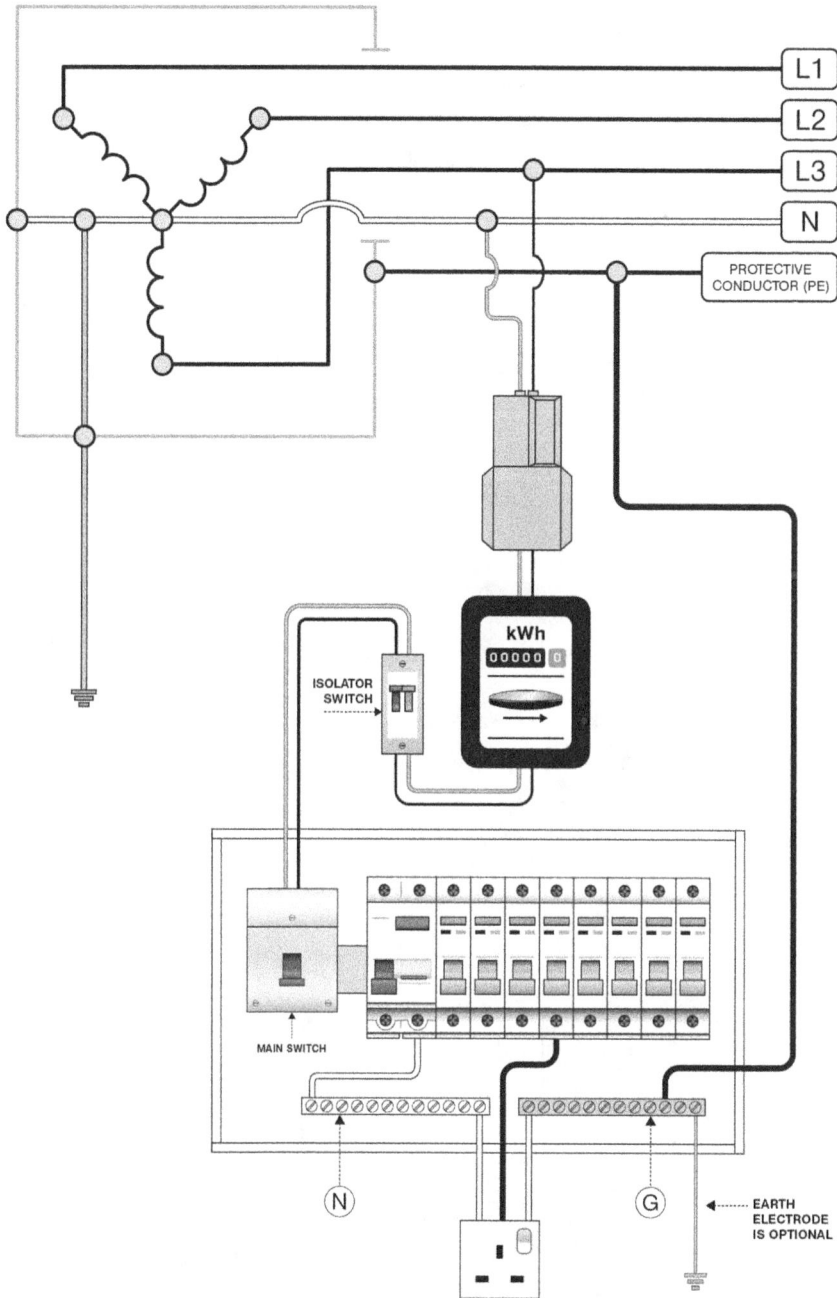

Figure 6.1 TN-S system.

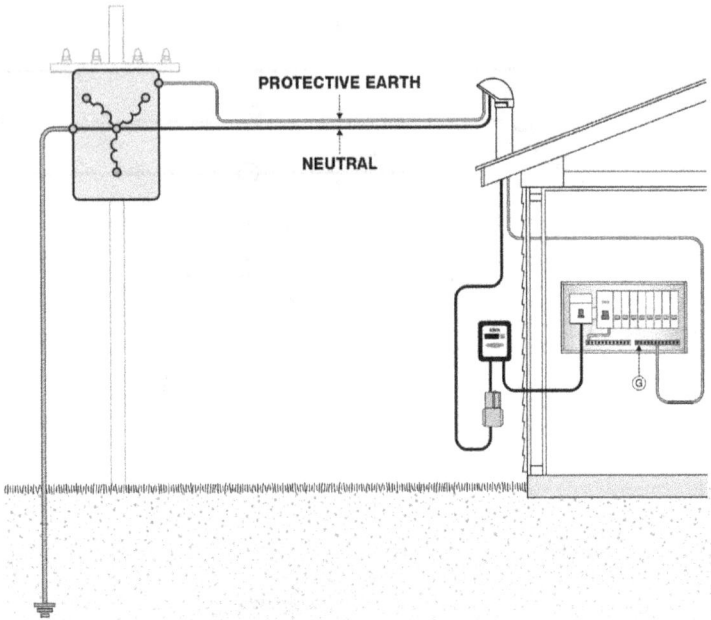

Figure 6.2 TN-S(2)/utility pole.

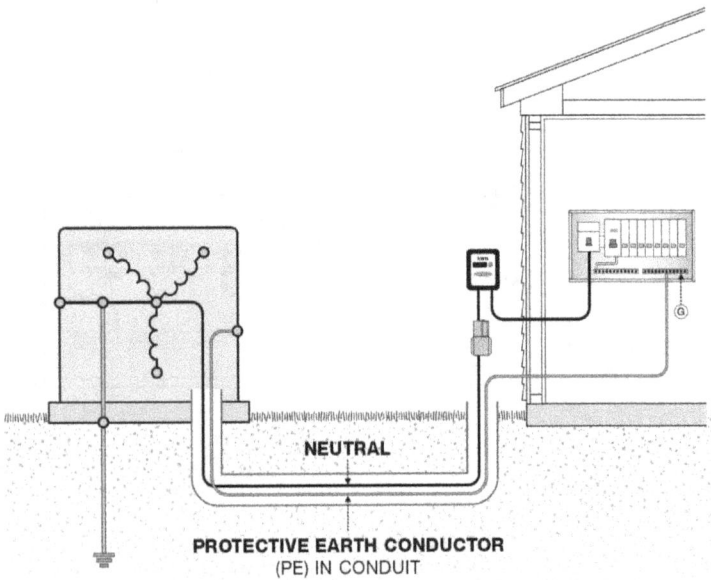

Figure 6.3 TN-S(2)/transformer.

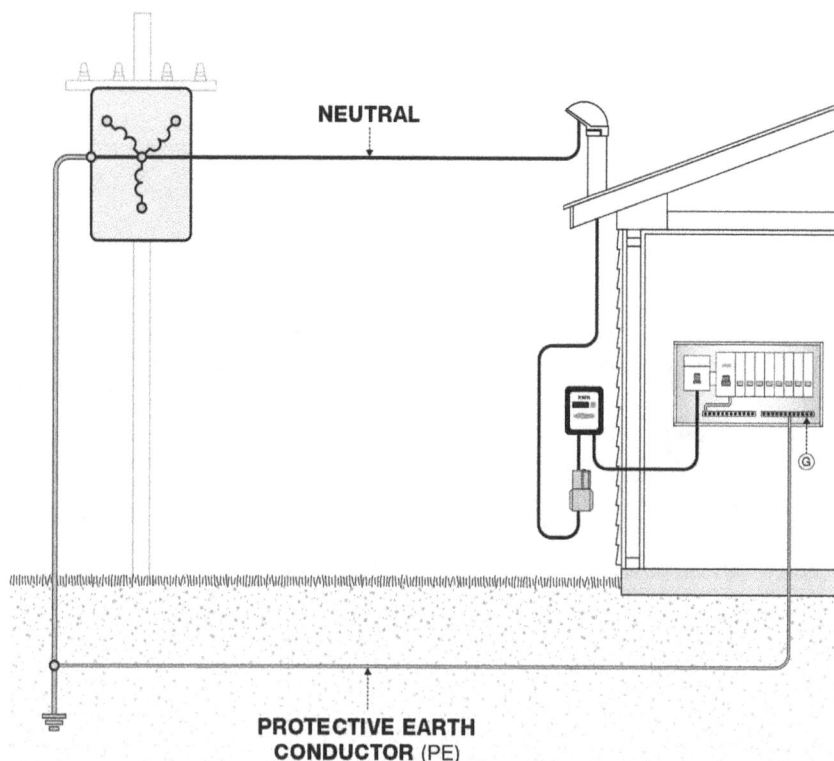

NEUTRAL

PROTECTIVE EARTH
CONDUCTOR (PE)

Figure 6.4 TN-S/utility pole.

Note that there is no neutral-to-ground bond at the distributor's cut-out.

An earth rod electrode (not shown) installed at the consumer's earth bar (marked with a circle G) is optional for TN-S systems, but highly recommended. In this case, the buried protective conductor will act as an effective earth electrode.

Figure 6.5 shows the same TN-S system illustrated in Fig. 6.1, but in a layout format. The line conductor has been removed from the illustration for clarity reasons, and only shows the neutral and protective conductor (protective earth).

In this case, the distributor (electric utility provider) is providing underground power to the installation, via three conductors to the installation: line wire (not shown for clarity), neutral wire, and a protective conductor wire that is buried and in direct contact with the earth.

Note that there is no neutral-to-ground bond at the distributor's cut-out.

An earth rod electrode (not shown) installed at the consumer's earth bar (marked with a circle G) is optional for TN-S systems, but highly recommended. In this case, the buried protective conductor will act as an effective earth electrode.

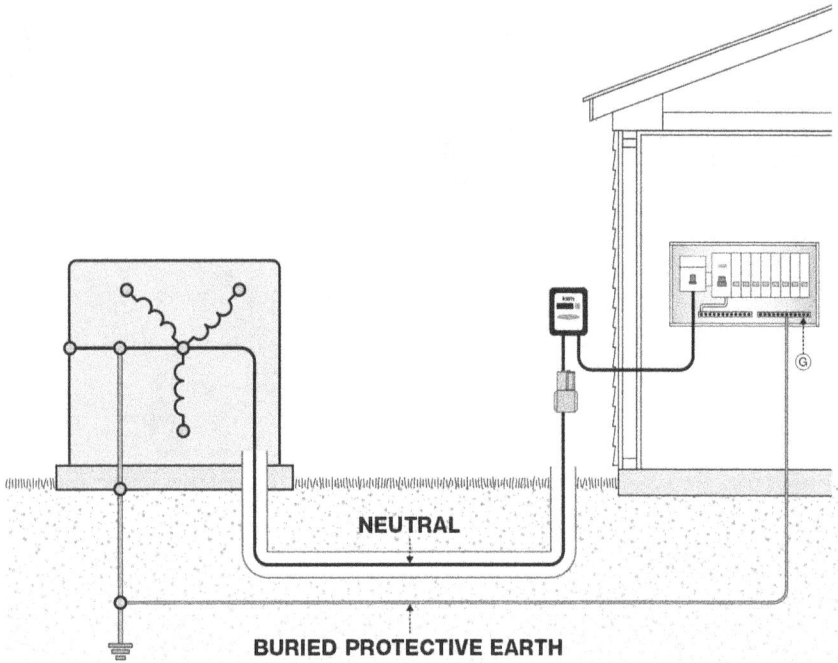

NEUTRAL

BURIED PROTECTIVE EARTH

Figure 6.5 TN-S/transformer.

TN-C-S SYSTEMS

In a TN-C-S system (the most common electrical system), the earthing electrode is connected to the star or midpoint of the power transformers winding, and only a single earthed conductor is provided to the main switchgear: a neutral conductor that also acts as the protective earth conductor. This combined (C) conductor is referred to as the PEN conductor. When the PEN conductor reaches the consumer's main switchgear, it is then separated into two individual conductor systems—the neutral and the protective conductor. In a TN-C-S system, there is a single connection between the neutral and PE conductors maintained at the consumer's switchgear. The PE must have a cross-sectional area of at least 16 mm².

A typical TN-C-S system is found when the incoming utility feed has two conductors, a hot and a PEN conductor. In a TN-C-S system, there are at two connections between the neutral and the protective earth conductors, one at the midpoint of the transformer and another one at the consumer's main switchgear.

The fault return path (fault-current loop) in a TN-C-S system is through the protective earth conductor up to the consumer's switchgear, and then through the PEN conductor to the transformer.

Earthing electrodes are not required to be installed at the consumer's main switchgear for TN-S and TN-C-S systems. However, there is no reason not to install these earthing electrodes, and there are in fact many reasons you should install them. A quick look through BS 7671 *Guidance Note 8* and you will find numerous illustrations showing such "means of earthing" to the MET. Additionally, certain countries may require you to install earthing electrodes at the consumer's switchgear.

The NEC and numerous international standards actually require a minimum of two earthing electrodes at the consumer's main switchgear for TN-S and TN-C-S systems.

Note: The NEC and some countries have requirements for specific resistances to ground of the earthing electrodes (such as 25 Ω or less), or they have maximum ground potential rise (GPR) values (such as 5000 V or less) of the earthing systems so that remote electrical monitoring equipment can properly detect an electrical fault occurring on a TN-C-S system.

TN-C-S ILLUSTRATIONS

Figure 6.6 shows a typical TN-C-S system from the transformer to the final circuit (socket-outlet). TN-C-S systems are almost certainly the most common electrical system in world (used in both IEC and NEC countries). While arguably not as safe as a TN-S system, they do have tried and proven record of accomplishment for safe operation around the world.

The important part that makes this a TN-C-S system, is that the distributor (electric utility provider) is only providing two conductors to the installation: line wire, and a combined neutral wire (N) and protective earth (PE) conductor, called a PEN conductor.

The PEN conductor is a current-carrying conductor and must be kept separated from the public as if it were a line conductor.

Note that there is a mandatory neutral-to-ground bond at the distributor's cut-out.

The earth rod electrode installed at the consumer's earth bar (marked with a circle G) is optional for TN-C-S systems, but highly recommended.

Figure 6.7 shows the same TN-C-S system illustrated in Fig. 6.6, but in a layout format. The line conductor has been removed from the illustration for clarity reasons, and only shows the neutral and PEN conductor.

In this case, the distributor (electric utility provider) is providing overhead power to the installation, via two conductors to the installation: line wire, and a PEN conductor [combined neutral wire (N) and protective earth (PE) conductor].

The PEN conductor is a current-carrying conductor and must be kept separated from the public as if it were a line conductor.

Note that there is a mandatory neutral-to-ground bond at the distributor's cut-out.

The earth rod electrode installed at the consumer's earth bar (marked with a circle G) is optional for TN-C-S systems, but highly recommended.

Figure 6.8 shows the same TN-C-S system illustrated in Fig. 6.6, but in a layout format. The line conductor has been removed from the illustration for clarity reasons, and only shows the neutral and PEN conductor.

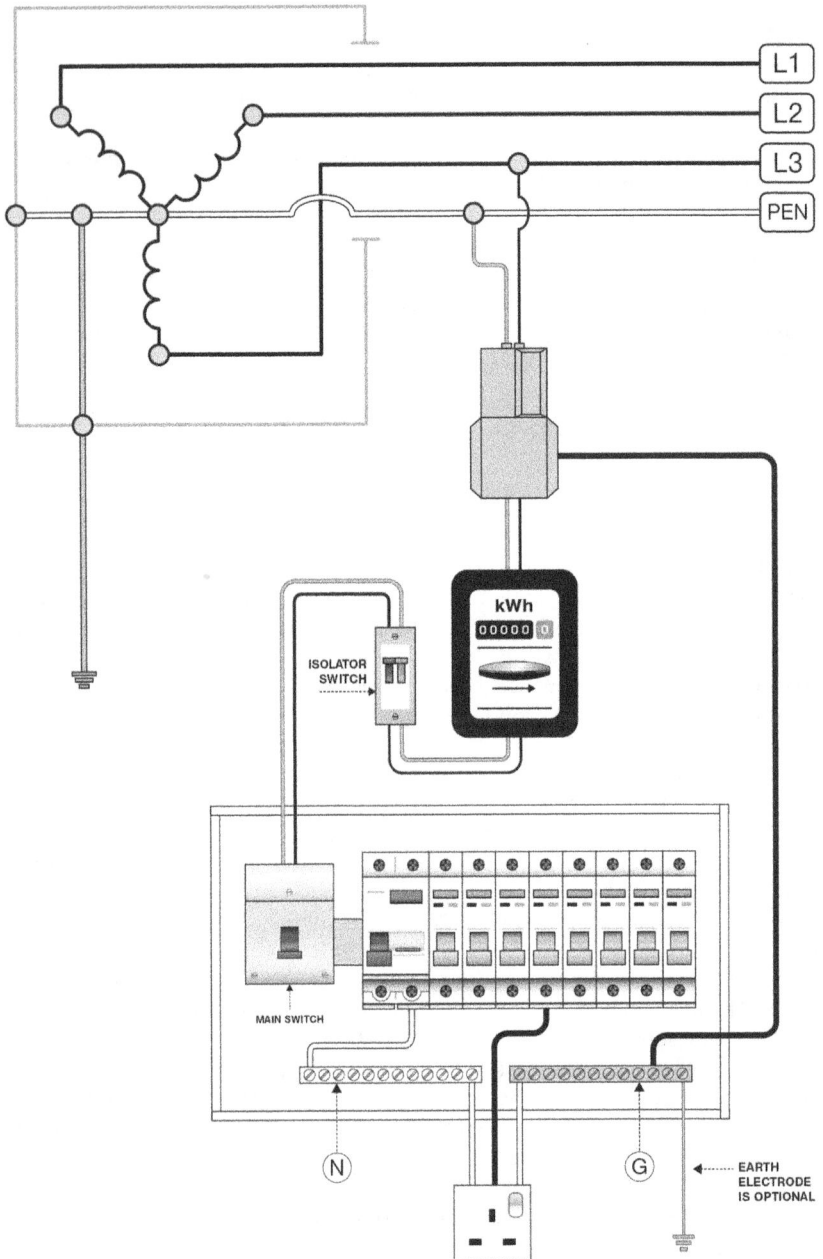

Figure 6.6 TN-C-S (PME) system.

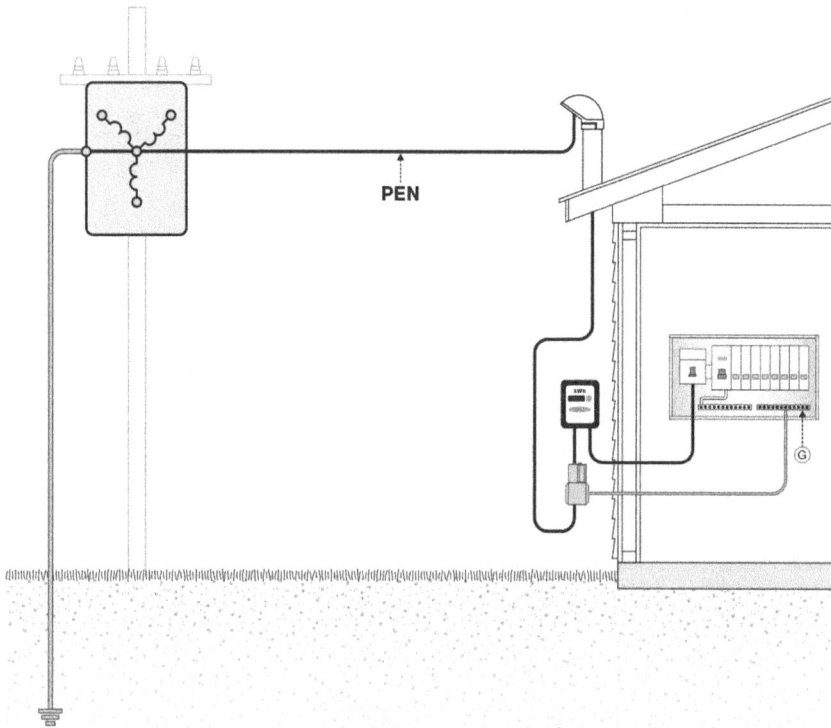

Figure 6.7 TN-C-S (PME)/utility pole.

In this case, the distributor (electric utility provider) is providing underground power to the installation, via two conductors to the installation: line wire, and a PEN conductor [combined neutral wire (N) and protective earth (PE) conductor].

The PEN conductor is a current-carrying conductor and must be kept separated from the public as if it were a line conductor.

Note that there is a mandatory neutral-to-ground bond at the distributor's cut-out.

The earth rod electrode installed at the consumer's earth bar (marked with a circle G) is optional for TN-C-S systems, but highly recommended.

IT SYSTEMS

In an IT system, there are no grounded conductors involved in the electrical system. IT systems typically utilize three-phase power only, and are not allowed to power any single or split (two-phase) phase systems. However, the chassis and exposed non-current-carrying metallic components are to bonded and earthed in accordance with BS 7671 Part 4 and BS 7671 Appendix 9.

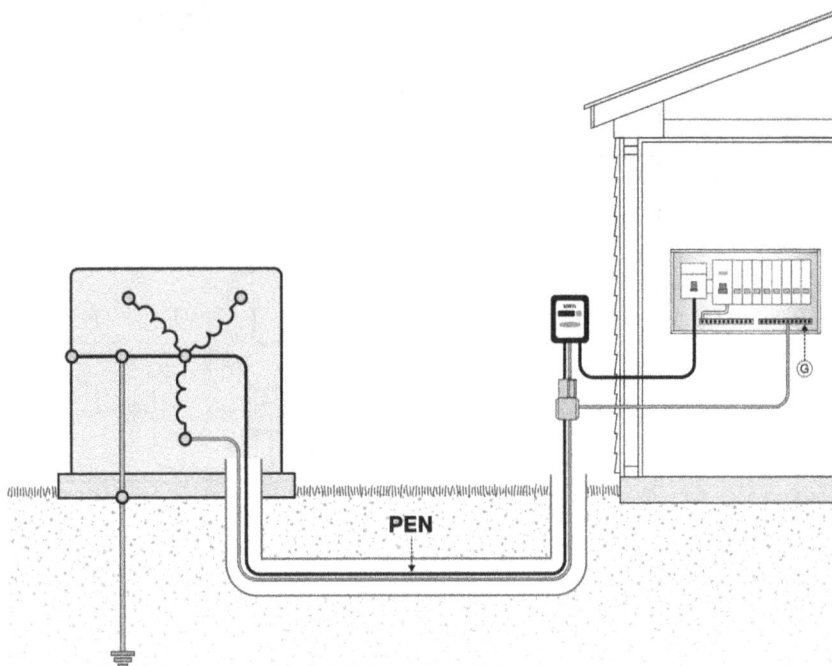

Figure 6.8 TN-C-S(PME)/transformer.

The fault return path (fault-current loop) in an IT system (an electrical fault to the exposed metal enclosure) is through the required earthing electrodes, through the earth itself, to the transformer. As the resistance of the earth is not known, special monitoring equipment as listed in BS 7671 Chapter 53 is required to be installed on IT systems.

Note: The NEC and some countries have requirements for specific resistances to ground of the earthing electrodes (such as 25 Ω or less), or they have maximum ground potential rise (GPR) values (such as 5000 V or less) of the earthing systems so that remote electrical monitoring equipment can properly detect an electrical fault occurring on an IT system.

TT SYSTEMS

In a TT system, the first required earthing electrode is connected to the star or midpoint of the power transformers winding, and only a single earthed conductors is provided to the main switchgear, a neutral conductor that also acts as the protective earth conductor. This combined (C) conductor is referred to as the PEN conductor. When the PEN conductor reaches the consumer's main switchgear, it is then separated into two individual conductor systems, the neutral and the protective earth conductor. In a TT system, the neutral and PE conductors are completely separated at the consumer's switchgear (no connection between the conductors). When this occurs, a second earthing electrode must be installed at the consumer's switchgear. The PE must have a cross-sectional area of at least 16 mm².

A typical TT system is found when the incoming utility feed has two conductors, a hot and a PEN conductor. In a TT system, there is only a single connection between the PEN conductor and earth found at the midpoint of the transformer. There is no (zero) connection made between the PE and neutral conductors at the consumer's main switchgear.

The fault return path (fault-current loop) in a TT system is through the required earthing electrode at the consumer's main switchgear, through the earth itself, to the transformer. As the resistance of the earth is not known, TT systems must be tested for resistance to earth of the consumer's earthing electrode in accordance with BS 7671 612.7 and 612.8.1 b, and for the earth loop impedance in accordance with BS 7671 612.8.1 b, 612.9, and Appendix 14.

Note: TT systems are forbidden under many electrical codes around the world as they use the earth (soil) as a conductor in the earth fault loop. As such, they are extremely dangerous as circuit breakers and fuses are not likely to function properly should a fault occur. BS 7671 should *not* allow the use of TT systems at any installation.

TT systems can easily be converted into a TN-C-S system simply by adding a jumper between the MET at the consumer's main switchgear and the incoming PEN conductor.

TT ILLUSTRATIONS

Figure 6.9 shows a typical TT system from the transformer to the final circuit (socket-outlet). TT systems are uncommon as they are without a doubt the most unsafe earthed electrical system possible. TT systems are so unsafe that they are actually illegal in every NEC-based country on the planet, and the IEC should ban TT systems as well.

The important part that makes this a TT system, is that the distributor (electric utility provider) is only providing two conductors to the installation: line wire, and a neutral wire (N).

In a TT system there is NO neutral-to-ground bond at the distributor's cut-out, which is what makes it so dangerous. A TT system relies on the earth (soil) as part of its fault-current path, where TN-S and TN-C-S systems use an all-metallic path.

The neutral conductor is a current-carrying conductor and must be kept separated from the public as if it were a line conductor.

The earth rod electrode installed at the consumer's earth bar (marked with a circle G) is mandatory for TT systems, and must have an earth resistance low enough to pass an earth loop impedance test (Z_s).

Figure 6.10 shows the same TT system illustrated in Fig. 6.9, but in a layout format. The line conductor has been removed from the illustration for clarity reasons, and only shows the neutral conductor.

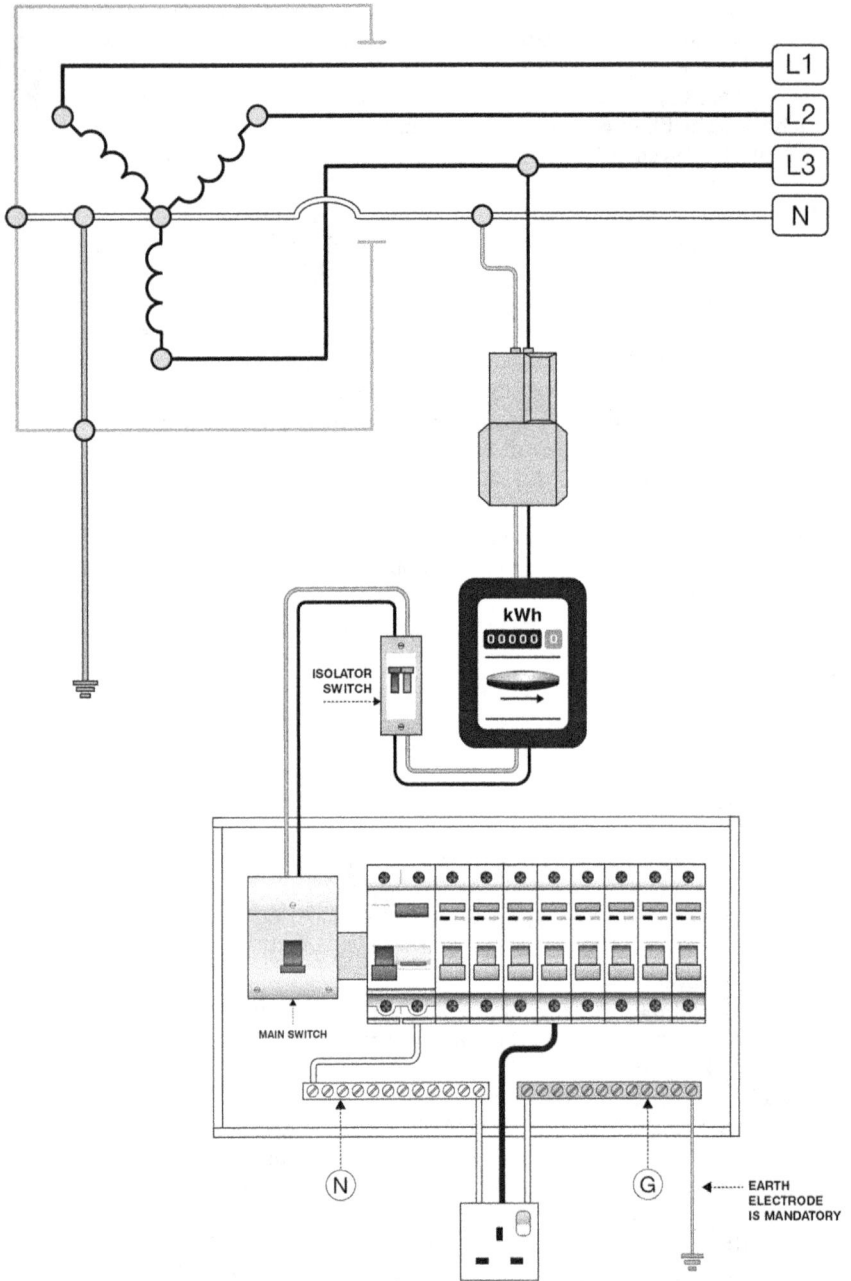

L1
L2
L3
N

ISOLATOR
SWITCH

kWh
00000 0

MAIN SWITCH

N

G ◄------- EARTH
ELECTRODE
IS MANDATORY

Figure 6.9 TT system.

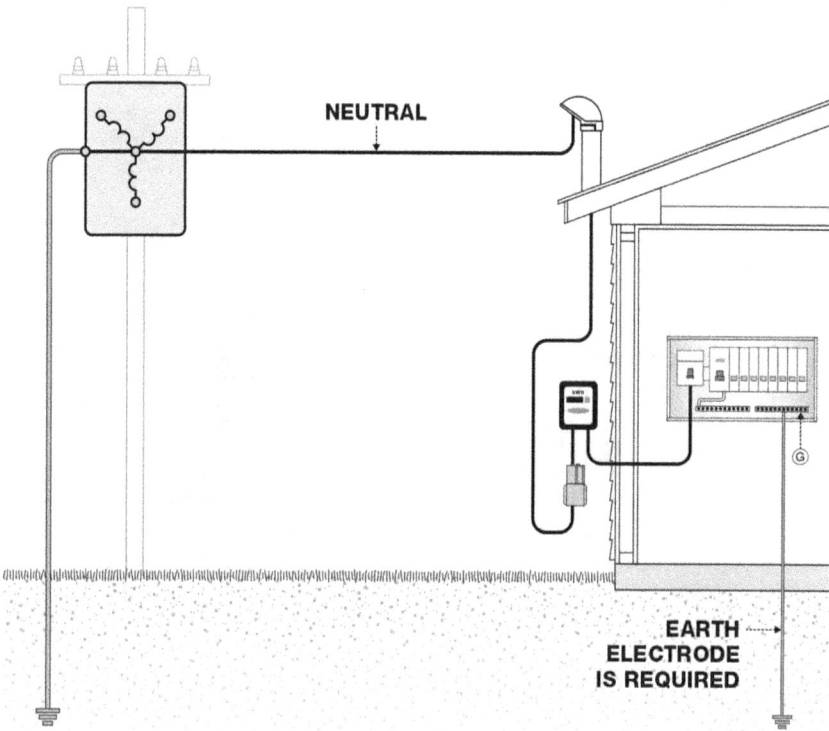

Figure 6.10 TT/utility pole.

In this case, the distributor (electric utility provider) is providing overhead power to the installation, via two conductors to the installation: line wire (not shown), and neutral wire (N).

In a TT system there is NO neutral-to-ground bond at the distributor's cut-out, which is what makes it so dangerous. A TT system relies on the earth (soil) as part of its fault-current path, where TN-S and TN-C-S systems use an all-metallic path.

The neutral conductor is a current-carrying conductor and must be kept separated from the public as if it were a line conductor.

The earth rod electrode installed at the consumer's earth bar (marked with a circle G) is mandatory for TT systems, and must have an earth resistance low enough to pass an earth loop impedance test (Z_s).

Figure 6.11 shows the same TT system illustrated in Fig. 6.9, but in a layout format. The line conductor has been removed from the illustration for clarity reasons, and only shows the neutral conductor.

In this case, the distributor (electric utility provider) is providing underground power to the installation, via two conductors to the installation: line wire (not shown), and neutral wire (N).

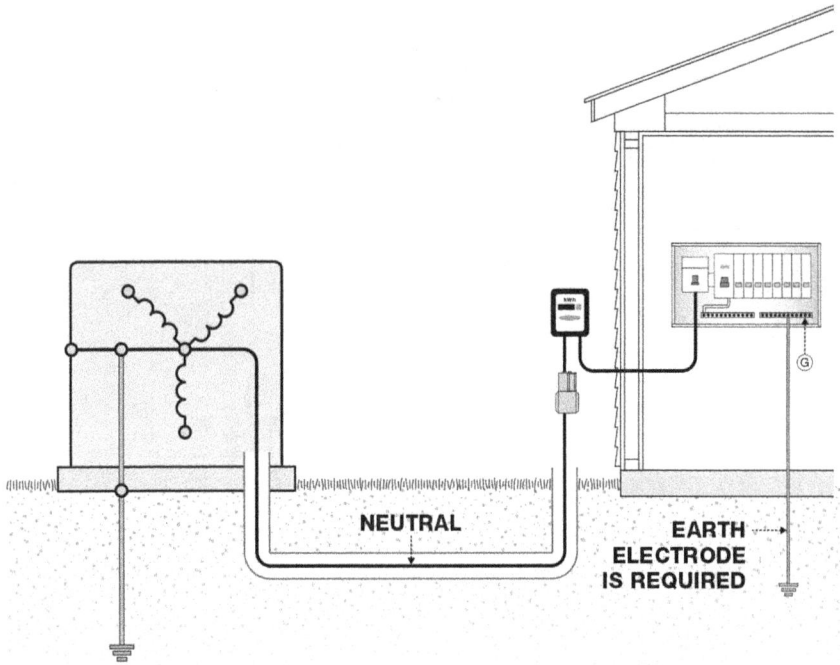

Figure 6.11 TT/transformer.

In a TT system there is *no* neutral-to-ground bond at the distributor's cut-out, which is what makes it so dangerous. A TT system relies on the earth (soil) as part of its fault-current path, where TN-S and TN-C-S systems use an all-metallic path.

The neutral conductor is a current-carrying conductor and must be kept separated from the public as if it were a line conductor.

The earth rod electrode installed at the consumer's earth bar (marked with a circle G) is mandatory for TT systems, and must have an earth resistance low enough to pass an earth loop impedance test (Z_s).

EARTH ELECTRODES

We find in 542.2 the physical requirements for earthing electrodes, required protection for the electrodes, and what can and cannot be used as an electrode.

For most installations, BS 7671 only requires an earthing electrode to be installed at TT system installations. For the most part, TN-S and TN-C-S installations are exempted from installing a dedicated earthing electrode. However, in some rural installations that are supplied by a protective multiple earthing (PME) arrangement, IET *Guidance Note 5* may require an earthing electrode installation to the consumer's MET, in order to mitigate the effects of a PEN conductor becoming an open-circuit.

It should be noted that there are no engineering reasons for not installing an earthing electrode at the consumer's MET for TN-S and TN-C-S installations. In fact, there are many positive reasons why you should always install an earthing electrode at every installation. It is simply that the Code does not mandate you to do so.

Interestingly, we do not find many instructions in either BS 7671 or BS 7430 on how earthing electrodes are to be installed, outside of the instruction in BS 7671 542.2.1 that earth electrodes shall be installed as such to withstand damage and take in to account the possibility that corrosion will increase the resistance over time. Many international standards, including the NEC have a variety of requirements for the installation of earthing electrodes including:

- Depth below grade that the electrode must be installed
- Length that the electrode must be in contact with the earth
- Type of connectors that must be used between the electrode and the earthing conductors (especially for buried connections)
- The minimum spacing between electrodes (see sphere of influence)
- The minimum resistance to earth of the electrode in certain cases
- The number of earthing electrodes

Fortunately, there is nothing in the BS 7671 or BS 7430 that prevents you from installing earthing electrodes at the consumer's service, or in following some common sense guidelines when installing earthing electrodes. Please refer to Chapter 9, the earthing section of this book, for more details on the selection and installation of various earthing electrodes.

The following electrodes are considered suitable electrodes that meet BS 7671 regulations, as found under 542.2.3:

- Earth rods or pipes
- Earth tapes or wires
- Earth plates
- Underground structural metal work embedded in foundations (steel rebar in concrete)
- Welded metal reinforcement of concrete (except pre-stresses concrete) embedded in the ground
- Lead sheathes and other metal coverings of cables, where not precluded by Regulation 542.2.5 (see below)
- Other suitable underground metalwork

In BS 7430:2011+A1:2015 we find a number of types of earthing electrode systems (arrangements) such as, triad or triangular arranged earth rods, mesh or earth grids, star arrangements, cruciform arrangements, hollow squares, and other various symmetrically arranged earth rods. All are considered valid under BS 7671.

We also find in BS 7430:2011+A1:2015 that we should have a preference for soil conditions when situations allow. This code tells us in BS 7430 9.2.1 that, when possible, our earthing systems should be placed in soil where the moisture content is continuously within the range of 15–20%. The standard tells us that following soil types are best, presented in order of preference:

1. Wet marshy ground
2. Clay, loamy soil, arable land, clayey soil, or loam mixed with small quantities of sand
3. Clay and loam mixed with varying proportions of sand, gravel, and stones
4. Dam and wet sand, peat

We are also told that where possible we are to avoid dry sand, gravel, chalk, limestone, whinstone, granite, very stony ground, and virgin rock. A location that is waterlogged in not necessary, unless the soil is sand or gravel. Sites with flowing water, either above or below the surface, should be avoided.

CONCRETE FOUNDATIONS, STRUCTURAL STEEL PILINGS, AND OTHER UNDERGROUND METAL WORKS

Part 4 of the Code mandates that you bond together all of the conductive metal objects in your structure together to form a single equipotential system, in order to protect people from various electrical hazards. The Code also allows you to use structural steel underground metal works, such as steel pilings, concrete encased steel rebar cage footings, concrete encased rebar foundations, and so on, as an earthing electrode.

We will see in the earthing conductor section below that there are basic size requirements of the underground steel works when we use it for this purpose. Mainly that the steel must have a cross-sectional area of at least 50 mm² when not protected from corrosion, or 16 mm² coated steel when protected from corrosion and protected from mechanical damage (such as being placed in concrete).

While you should most definitely bond the steel rebar components to your earthing system, I am not a fan of using it as part of your fault-current loop. Concrete may crack open should the electrode take an excessive electrical fault, as the water inherent in the concrete can be superheated by the current and turned in to steam, thus expanding and destroying the concrete. We are warned about this specifically in BS 7430:2011+A1:2015 Section 9.5.8.6.

Additionally, we are told in this section that continuous electrical currents can cause significant corrosion and that steel will expand to many times its volume when it corrodes, thus cracking and destroying the integrity of the concrete.

What we do not see stated clearly in the standards, and it should be obvious, is that steel rebar that is coated in plastic, or when an insulating vapour barrier is installed between the concrete and the earth, then these structural steel members are no longer in contact with the earth. You have placed an insulator that will prevent current from flowing through the steel members in to the earth. When this occurs, you may not consider these objects as an earthing electrode, however, you do still need to bond them to the protective earthing system as required under Part 4 of the Code.

SOIL TREATMENT FOR EARTHING ELECTRODES

BS 7430:2011+A1:2015 Article 9.2.2 instructs us to use a soil treatment around our earthing electrodes when installing them in high-resistivity locations or on rocky ground. We are told *not* to use coke breeze (iron slag) due to its highly corrosive nature. This corrosiveness is due primarily to the differential between the nobility of the metal in the earthing electrodes and the nobility of the coke breeze, which makes the earthing electrode sacrificial to the coke breeze.

We are told that there are commercially available products that can be purchased, such as conductive concrete that can be used as a soil treatment. The problem is that nearly every product in the market today, including conductive concrete, uses coke breeze in its formulation. Additionally, concrete may crack open should the electrode take an excessive electrical fault, as the water inherent in the concrete will be superheated by the current and turned in to steam, thus expanding and destroying the concrete. Conductive concrete products should be avoided for this reason.

A few manufacturers claim to add an "enzyme" to the coke breeze to prevent its corrosive nature. This claim has not been proven, and these manufacturers have not yet presented peer-reviewed data on the effectiveness of these "enzymes." Coke breeze products with "enzymes" should not be used.

This leaves us with only one true option for soil enhancement and that is bentonite clay. Bentonite clay has the distinct advantage of being both hygroscopic and slightly caustic (has a pH that is slightly on the base side), which means that it absorbs water and will protect the electrode from corrosion. However, again we find in BS 7430 9.2.2 to be cautious of soil treatments that may shrink, which is exactly what will happen with bentonite clay should it become dry. Which means that you can use bentonite clay if you have a reliable water supply for the clay. If you do not have a reliable water supply, you are better off using native soil than any other known back fill.

You will find a great number of formulas for calculating the resistance to earth of various earthing electrodes in Chapter 8, the earthing calculations section of this book.

EARTHING ELECTRODES FOR TT SYSTEMS

BS 7671 mandates the installation of an earthing electrode at the consumer's MET for every TT system installation. This earth electrode must be measured (tested) under a number of conditions.

If a RCD is being used to protect a TT installation (and there should be one!), then the earthing electrode must be tested using the "loop impedance method." The results of the loop impedance test must show a measured result of less than 200 ohms. See Chapter 7 in this book on Part 6 for more information on loop impedance tests.

Additionally, the earth fault loop impedance must either be directly measured (Z_s), or calculated by adding the results of the continuity test to the resistance results of an earth electrode measurement. While there is no set of maximum resistance requirement for the earth electrode itself, there is a maximum earth fault loop impedance, which is again 200 ohms. See the chapter on Part 6 for more information on earth fault loop impedance measurements.

EARTHING CONDUCTORS

The earthing conductor, where installed, is the conductor that connects the MET of an installation to an earthing electrode of other means of earthing. We find in BS 7671 Section 542.3.1 that *every* earthing conductor must comply with Section 543,

Protective Conductors, and where PME conditions apply, the earthing conductors must additionally meet the requirements found in 544.1.1 for cross-sectional area. **Earthing Conductor.** A protective conductor connecting the main earthing terminal of an installation to an earth electrode or to other means of earthing.

Generally, as we see in BS 7430 Section 6.7, earthing conductors are sized the same way that all other protective conductors are sized, with the exception that we have a minimum cross-sectional area requirement that we cannot go under. This means that the earthing conductor must be selected based upon a calculation (formula) plus a number of tables found within Part 5 of BS 7671.

We find a much simpler version of earthing conductors and main protective bonding conductors presented in Section 4.4 of the IET *On-Site Guide* for BS 7671:2011+A3:2015. However, the following table does not negate the need to confirm that our earthing conductor meets the requirements for all protective conductors (see protective conductor section below).

The following tables, also from the IET *On-Site Guide* for BS 7671:2011+A3:2015, shows the minimum size of the buried earthing conductors based on the type of materials used. Note that steel is generally expected to be the structural steel of the building that is in contact with the earth, such as a structural piling or concrete encased steel rebar in a foundation.

AUTHOR'S NOTE ON THE SELECTION OF EARTHING CONDUCTORS

So what is the bottom line when it comes to properly sizing the earthing conductor? First of all, you have to do the formula (see protective conductor section) to make sure that the earthing conductor can handle the fault currents. There is just no getting around it. However, you can do a few sanity checks:

1. The incoming neutral and line conductors should be the same size or possibly the neutral will be a size or two larger. If the neutral is smaller, call the distributor and find out why, because there is almost certainly a problem.
2. Base the size of your earthing conductor on the size of the incoming neutral.
3. Your earthing conductor should be copper. Even if you are connecting to steel rebar in your concrete foundation.
4. The earthing conductor should be at least 25 mm², except under very restricted conditions.

Honestly, why would you want to go any smaller than 25 mm² anyhow? It's really not worth your time (or money) with all the hassle it would take with the inspectors to use a smaller size conductor or to use aluminium conductors. So just use 25 mm² copper as your starting point and increase the size from there.

EARTHING AND BONDING

In IET *On-Site Guide* for BS 7671:2011+A3:2015 Chapter 4, we find an entire section dedicated to earthing and bonding. One of the main parts of the chapter is to

Minimum Earthing and Bonding Conductor Sizing for TN-S and TN-C-S Systems

	mm² 4	mm² 6	mm² 10	mm² 16	mm² 25	mm² 35	mm² 50	mm² 70	mm² 95	mm² 150	mm² 150+
Size of the Line or Neutral Conductor Supplying Power (from a PME source)											
Minimum required size of Earthing Conductor for not buried or buried and protected against corrosion and mechanical damage*	6¹	6¹	10¹	16	16	16	25	35	—	—	—
Required size of Main Protective Bonding Conductor	6¹	6¹	6¹	10¹	10¹	10¹	16	25	25	35	50
Required size of Main Protective Bonding Conductor for PME Supplies (TN-C-S)	10¹	10¹	10¹	10¹	10¹	10¹	16	25	25	35	50

See BS 7671 Section 542.3 and 543.1 (Minimum required size of Earthing Conductor row)

See BS 7671 Section 544.1.1 (Required size of Main Protective Bonding Conductor row)

See BS 7671 Table 54.8 (Required size of Main Protective Bonding Conductor for PME Supplies row)

*Earthing conductors must be calculated to handle the fault currents on the system in accordance with BS 7671 Section 543 Protective Conductors

Note 1: Per BS 7671 Section 543.2.4, protective conductors of 10 mm² cross-sectional area or less shall be copper

Note 2: Per BS 7671 Table 54.7, a 16 mm² minimum sized earthing conductor may be required by the distributor at the origin of the supply

Note 3: See Table in this book for Minimum Buried Earthing Conductor Sizes for additional minimum sizing requirements

Note 4: BS 7430 Section 9.7 states that only copper may be used for earthing conductors that come in contact with the soil or in damp environments

71

Minimum Buried Earthing Conductor Sizes for TN-S and TN-C-S Systems

Copper	Not protected against corrosion	25 mm^2
Steel	Not protected against corrosion	50 mm^2
Copper	Protected from corrosion but NOT protected from mechanical damage	16 mm^2
Steel	Protected from corrosion but NOT protected from mechanical damage	16 mm^2

See 542.3.1 and Table 54.1 in BS 7671

Minimum Earthing Conductor Cross-Sectional Area for TT Systems

Buried Copper	Unprotected	25 mm^2
Buried Copper	Protected against corrosion	16 mm^2
Buried Copper	Protected from corrosion and from mechanical damage	2.5 mm^2
Not Buried Copper	Unprotected	4 mm^2
Not Buried Copper	Protected against corrosion	4 mm^2
Not Buried Copper	Protected from corrosion and from mechanical damage	2.5 mm^2

Note 1: The corrosion protection is assumed to be a sheath
Note 2: The main protective bonding conductors for TT Systems shall have
a cross-sectional are of not less than half that required for the earthing conductor and not less than 6 mm^2

remind us of BS 7671 Section 411.3.1.2, which defines and gives us examples of various extraneous conductive parts:
- Metallic installation pipes
- Metallic installation gas pipes
- Other installation pipework such as heating oil pipes
- Building structural steelwork that rises from the ground
- Lightning protection systems as required under BS EN 62305

In the IET *On-Site Guide* for BS 7671:2011+A3:2015 we find Table 4.4(i) which gives us information regarding the cross-sectional area requirements for our supplementary bonding conductors.

PROTECTIVE CONDUCTORS

In BS 7671 Section 543.1.1 we are given a choice of how we would like to select the size (cross-sectional area) of our protective conductors. Either we can calculate the size in accordance with 543.1.3, or we can simply select the size in accordance with 543.1.4. We must calculate the size if the earth fault current is

Table 4.4 (i) Minimum Cross-Sectional Area or Supplementary Bonding Conductor

Size of Circuit Protective Conductor (mm²)	Exposed-Conductive-Part to Extraneous-Conductive Part		Exposed-Conductive-Part to Exposed-Conductive-Part		Extraneous-Conductive-Part to Extraneous-Conductive Part*	
	Mechanically Protected	NOT Mechanically Protected	Mechanically Protected	NOT Mechanically Protected	Mechanically Protected	NOT Mechanically Protected
Column #	A	B	C	D	E	F
1.0	1.0	4.0	1.0	4.0	2.5	4.0
1.5	1.0	4.0	1.5	4.0	2.5	4.0
2.5	1.5	4.0	2.5	4.0	2.5	4.0
4.0	2.5	4.0	4.0	4.0	2.5	4.0
6.0	4.0	4.0	6.0	6.0	2.5	4.0
10.0	6.0	6.0	10.0	10.0	2.5	4.0
16.0	10.0	10.0	16.0	16.0	2.5	4.0

*Per 544.2.3, if one of the extraneous-conductive-parts is connected to an exposed-conductive-part, the bonding conductor must be no smaller than that required in column A or B

Table 54.7

Cross-sectional area of line conductor (S)	Line conductor and protective conductor are same material	Line conductor and protective conductor are NOT same material
mm²	mm²	mm²
S is less than 16 mm² $S \leq 16\,mm^2$	S	$\dfrac{k_1}{k_2} \times S$
S is greater than 16 mm² and less than or equal to 35 mm² $16 < S \leq 35\,mm^2$	16	$\dfrac{k_1}{k_2} \times 16$
S is greater than 35 mm² $S > 35\,mm^2$	S/2	$\dfrac{k_1}{k_2} \times \dfrac{S}{2}$

Where:
    k_1 = the value of k from Table 43.1
    k_2 = the value of k from Tables 54.2 to 54.6

expected to be less than the short-circuit current and the protective conductor meets the following:

- It is *not* an integral part of a cable
- It is *not* formed by conduit, ducting, trunking, and so on.
- It is *not* contained in an enclosure formed by a wiring system

If you do not need to calculate the size of your protective conductor as allowed under 543.1.4, Table 54.7 can be used to determine the required sizing. This table gives you two options, either the protective conductor in question will be made from the same material as the line conductor, or it will not be made of the same material.

If the materials of the line conductor and the PE will be the same, then the first column gives us a very simple method of determining the required size of our PE. On the other hand, if the materials will be different, such as the incoming line conductor is going to be aluminium and our PE must be copper (as aluminium may not come in contact with the earth according to BS 7430 Section 9.7), then we must use column two. As you can see in Table 54.7, in order to use this column, we must have two factors: k^1 and k^2 from Tables 54.2 to 54.6. Fortunately, these are the same factors come from the same tables needed if we are going to calculate the conductor size according to 543.1.3.

Table 54.7 is the primary table for determining the size of every protective earthing conductor in your electrical system, as governed under BS 7671 Sections 543.1.1 and 543.1.4.

Table 43.1 is used to determine the k^1 factor required in Table 54.7. Table 43.1 is also used for the calculation of fault-current protective devices under BS 7671 Section 434.5.

Tables 54.2 to 54.6 provide the k_2 factor needed in Table 54.7.

Table 43.1

Values of k for common materials, for calculation of the effects of the fault current for disconnection times up to 5 seconds

	Conductor Insulation							
	Thermoplastic				Thermosetting		Mineral insulated	
	Conductor cross-sectional area						Thermoplastic sheath	Bare (unsheathed)
	≤300 mm²	>300 mm²	≤300 mm²	>300 mm²				
Initial Temperature	90°C		70°C		90°C	60°C	70°C	105°C
Final Temperature	160°C	140°C	160°C	140°C	250°C	200°C	160°C	250°C
Copper Conductor	k = 100	k = 86	k = 115	k = 103	k = 143	k = 141	k = 115	k = 135/115[a]
Aluminium Conductor	k = 66	k = 57	k = 76	k = 68	k = 94	k = 93		
Tin soldered joints in copper conductors	k = 100	k = 86	k = 115	k = 103	k = 100	k = 122	k = 115	

[a] = This value shall be used for bare cables exposed to touch

Table 54.2

Values for k for insulated protective conductor not incorporated in a cable and not bunched with cables, or for separate bare protective conductor in contact with cable covering but not bunched with cables, where assumed initial temperature is 30°C.

Material of conductor	Insulation of protective conductor or cable covering		
	70°C Thermoplastic	90°C Thermoplastic	90°C Thermosetting
Copper	143/133*	143/133*	176
Aluminium	95/88*	95/88*	116
Steel	52	52	64
Assumed Initial Temperature	30°C	30°C	30°C
Final Temperature	160°C/140°C*	160°C/140°C*	250°C

*Above 300 mm²

Table 54.3

Values for k for insulated protective conductor incorporated in a cable or bunched with cables, where the assumed initial temperature is 70°C or greater

Material of conductor	Insulation of protective conductor or cable covering		
	70°C Thermoplastic	90°C Thermoplastic	90°C Thermosetting
Copper	115/103*	100/86*	143
Aluminium	76/68*	66/57*	94
Assumed Initial Temperature	70°C	90°C	90°C
Final Temperature	160°C/140°C*	160°C/140°C*	250°C

*Above 300 mm²

Table 54.4

Values for k for insulated protective conductor as the sheath or armour of a cable as the protective conductor

Material of conductor	Insulation of protective conductor or cable covering		
	70°C Thermoplastic	90°C Thermoplastic	90°C Thermosetting
Aluminium	93	85	85
Steel	51	46	46
Lead	26	23	23
Assumed Initial Temperature	60°C	80°C	80°C
Final Temperature	200°C	200°C	200°C

Table 54.5

Values for k for steel conduit, ducting and trunking as the protective conductor

Material of conductor	Insulation of protective conductor or cable covering		
	70°C Thermoplastic	90°C Thermoplastic	90°C Thermosetting
Steel conduit, ducting and trunking	47	44	58
Assumed Initial Temperature	50°C	60°C	60°C
Final Temperature	160°C	160°C	250°C

Table 54.6

Values for k for bare conductor where there is no risk of damage to any neighbouring material by the temperature indicated

Material of conductor	Condition		
	Visible and in restricted area	Normal conditions	Fire risk
Copper	228	159	138
Aluminium	125	105	91
Steel	82	58	50
Assumed Initial Temperature	30°C	30°C	30°C
Final Temperature			
Copper	500°C	200°C	150°C
Aluminium conductor	300°C	200°C	150°C
Steel conductor	500°C	200°C	150°C

If we find that we do need to calculate the size of our protective conductors, there are a number of factors that must be considered when selecting the size of the earthing conductors for your electrical system. Primarily, we need to know:

- The size of the incoming line conductor
- The fault current
- Any environmental issues at the site, such as extreme temperature, physical hazards, chemical hazards, and so on
- How we plan to route and connect the earthing conductor to the earthing electrode from the MET (i.e., in conduit, above-grade or below-grade, etc.)
- What type of materials will be used for the earthing conductor (copper, aluminium, steel, copper-aluminium mix, etc.)

Finding the size of the incoming line conductor is easy, but how does one get the fault current? BS 7430:2011+A1:2015 tells us that Regulations 28 and 29 of the Electricity Safety, Quality, and Continuity Regulations (ESQCR) require the distributor to provide certain information as follows (the exact text here is taken from the ESQCR):

> 28. A distributor shall provide, in respect of any existing or proposed consumer's installation which is connected or is to be connected to his network, to any person who can show a reasonable cause for requiring the information, a written statement of
>   (a) the maximum prospective short-circuit current at the supply terminals;
>   (b) for low-voltage connections, the maximum earth loop impedance of the earth fault path outside the installation;
>   (c) the type and rating of the distributor's protective device or devices nearest to the supply terminals;
>   (d) the type of earthing system applicable to the connection; and
>   (e) the information specified in Regulation 27(1), which apply, or will apply, to that installation.

In other words, the electric company (distributor) must provide you with the fault current.

This leads us to the formula found in BS 7671 Section 543.1.3, which tells us that the cross-sectional area of our earthing conductor shall not be less than what is calculated using the following:

$$S = \frac{\sqrt{I^2 t}}{k}$$

where

S = the nominal cross-sectional area of the conductor in mm^2

I = the fault current value in amperes rms (provided by the distributor and assuming a negligible impedance from the earth)

t = operating time of the protective device (clearing time) for the fault (I) in seconds, as provided by the distributor

k = a factor taking into account the resistivity, temperature coefficients, heat capacity, and so on of the conductor's material properties. Appropriate initial and final temperatures must also be accounted for. See Tables 54.2 to 54.6 for k factor information.

Circuit Protective Conductor (CPC).   A protective conductor connecting exposed conductive parts of equipment to the main earthing terminal.

TYPES OF PROTECTIVE CONDUCTORS

In BS 7671 543.2.1 we find that a protective conductor (PE) may consist of any single or any combination of the following:
1. Single-core cable
2. A conductor in a cable
3. An insulated or bare conductor in a common enclosure with insulated live conductors
4. A bare of insulated conductor that is fixed
5. A sheath, screen, or armouring of a cable or other type of metal covering

6. The metal conduit, metal enclosure, metal cable management system, or other metallic electrically continuous support system for conductors

7. Other extraneous conductive part that is in compliance with 543.2.6

In 543.2.4, we find that for items 1 through 4 above, that have a cross-sectional area of less than 10 mm², shall only be made of copper.

We also find in 543.2.2 and 543.2.5 that we can use the following as a protective conductor for the associated circuit, given certain requirements are met:

- Metal enclosure or frame of a low-voltage switchgear—must meet a, b, and c below
- Metal enclosure or frame of a control gear assembly—must meet a, b, and c below
- Metal enclosure or frame of a busbar trunking system—must meet a, b, and c below
- Metal sheath (bare or insulated) of a cable—must meet a and b below
- Metal sheath of a mineral insulated cable—must meet a and b below
- Trunking or ducting for electrical purposes—must meet a and b below
- Metal conduit—must meet a and b below

In order for the above to qualify as a PE, all three of the following conditions that must be met:

a. The electrical continuity of the metal enclosure or frame must be assured, from the way it is constructed or other suitable connection, in such a way as to be protected from chemical, electrochemical, and/or mechanical deterioration

b. The metal enclosure used as a PE must have a cross-sectional area equal to that resulting from 543.1. If you are a manufacturer of an enclosure system, this may also be verified by test in accordance with the BS EN 61439 series of standards

c. The metal enclosure or frame must permit the connection of other protective conductors at every predetermined tap-off point

If the protective conductor is made up of metal conduit, trunking, ducting, or the metal sheath of a cable (including armoured cable), you have some additional requirements. Namely, that a separate PE is required to be installed to an earthing terminal incorporated in the associated box or other enclosure, as required under 543.2.7.

If you are dealing with a ring circuit, the PE for every ring final circuit must be run in the form of a ring having both ends connected to the earthing terminal at the origin of the circuit. Unless, the PE is formed by the metal covering or enclosure, in which case the metal covering or enclosure must contain all of the conductor of the ring circuit.

If your protective conductor is an extraneous conductive part, we find in 543.2.2 that certain requirements must be met:

a. The electrical continuity of the extraneous conductive part must be assured, from the way it is constructed or other suitable connection, in such a way as to be protected from chemical, electrochemical, and/or mechanical deterioration.

b. The extraneous conductive part used as a PE must have a cross-sectional area equal to that resulting from 543.1.

c. Precautions shall be taken to prevent its removal, unless compensatory measures have already been provided.

d. The extraneous conductive-part has been considered for such a use and, if necessary, suitably adapted.

A few items may *not* be used as a protective conductor according to BS 7671 Sections 543.2.3 and 543.2.10:

- Gas pipe
- Oil pipe
- Flexible conduit
- Pliable conduit
- Support wires
- Flexible metallic parts
- Constructional parts subjected to mechanical stress during normal operating service
- Separate metal enclosure(s) for the cable or circuit—additionally, this may *not* be used as a PEN conductor

PRESERVATION OF ELECTRICAL CONTINUITY OF PROTECTIVE CONDUCTORS

The Code asks us to provide protection and electrical continuity for our protective conductors under a number of different situations. As we see in 543.3.1, the main requirement is to ensure that our PE conductors are suitably protected from mechanical and chemical deterioration, along with electrodynamic effects.

Note: Apparently, the Code only uses the term "electrodynamic" in this section and asks us in the Index section of BS 7671 to reference "Electromechanical" and "Electromagnetic." From here, we are further referenced to the "Electromagnetic Compatibility Regulations 2006, Appendix 2 section 12," and to BS 7671 Sections 444, 521.5, 332.2, 521.5.201, 542.1.3.1(ii), and 332.

We also see in 543.3.2 that every connection and joint of the PE must be accessible for inspection and maintenance and testing, except as allowed under 526.3. BS 7671 Section 526.3 allows joints designed to be buried underground, compound-filled or encapsulated joints, welded joints, soldered joints, brazed joints, appropriate compression joints, maintenance-free joints, and other types of joints as listed, to be excluded from the accessibility rule.

Protective conductors that have a cross-sectional area of 6 mm² or less must be protected throughout its routing. This protective covering must be rated to provide the insulation of a single-core non-sheathed cable of appropriate size with a voltage rating of 450/750 V. This rule found in 543.3.201 does not include the PE forming part of a multicore cable, or when metal conduit, metallic cable management system, or other electrical enclosure that has been made electrically continuous. See the BS EN 60684 Series of standards for more information.

When using metallic conduit as the PE, every joint must be made electrically and mechanically continuous, as required under 543.3.6.

It is allowable under 543.3.3 to have intentional disconnection joints that are intended for testing purposes in the protective conductor circuit.

There are some things that we cannot do with our protective conductor. In 543.3.3, 543.3.4, and 543.3.5, we find that inserting a switching device in series with the PE is prohibited, except as allowed under 537.1.5 (such as an automatic transfer switch for an alternate power source). Also, multipole-linked switching or plug-in devices must make the PE connection prior to making the line or neutral connection. This is sometimes referred to as make-first-break-last technology. Where earth-monitoring devices are used, they may not be installed in series with the PE conductor (see BS 4444).

COMBINED PROTECTIVE AND NEUTRAL CONDUCTORS

The overriding principle of the PEN conductor is that it may *not* be used within an installation except under a few circumstances. In fact, in Great Britain Regulation 8(4) of the Electricity Safety, Quality, and Continuity Regulations 2002 prohibits the use of PEN conductors within a consumer's installation. These regulations are so serious, that if you feel you have an applicable use for a PEN conductor within an installation, we recommend you consult with an electrical engineer who is thoroughly versed in BS 7671 Section 543.3.

A PEN conductor must be made up of a copper conductor that has a cross-sectional area of 10 mm^2 or more, or an aluminium conductor that has a cross-sectional area of 16 mm^2 or more, must be for a fixed installation, and be a conductor of a cable not subject to flexing. A PEN conductor may not be supplied through an RCD. See BS 7671 Section 543.4.201.

EARTHING ARRANGEMENTS FOR COMBINED PROTECTIVE AND FUNCTIONAL PURPOSES

You must make protective measures the precedence over functional purposes when earthing according to BS 7671 Section 543.5.1.

EARTHING ARRANGEMENTS FOR PROTECTIVE PURPOSES

The protective conductor for an associated overcurrent device (fuse, circuit breaker, etc.) used for fault protection, must be incorporated into the same wiring system as the associated line conductors, or in their immediate proximity. See BS 7671 Section 543.6.1.

EARTHING REQUIREMENTS FOR THE INSTALLATION OF EQUIPMENT HAVING HIGH PROTECTIVE CONDUCTOR CURRENTS

So what is high level of current according to the Code? In BS 7671 Section 543.7 we find that for our non-current-carrying protective conductors there are three recognized levels of current that may be present on our PE, during normal operating conditions:

- 0 to less than 3.5 mA
- 3.5 to 10 mA
- Greater than 10 mA

Apparently, there are no additional requirements for PE conductors that have less than 3.5 mA of current on them during normal operations, as the Code makes no further mention of this condition. And why should it? This is exactly what we would hope to find!

Now, if your PE conductor has between 3.5 and 10 mA of current on it during normal operating conditions, we find in 543.7.1.201 that the equipment is to have a permanent electrical installation without the use of plug and socket-outlet connections. However, if your plug and socket-outlet complies with BS EN 60309-2, then you may run equipment with PE currents between 3.5 and 10 mA during normal operating conditions.

Now, if your equipment is likely to have, or does in fact have, PE current levels exceeding 10 mA, you have a series of requirements that must be met. Namely, you will have requirements regarding the supply, and requirements regarding the circuit, which we will discuss below. In addition (see 543.7.1.205), you must label the distribution board indicating those circuits that have high PE current levels (the Code does not make it clear if "high" current in this case means current levels in excess of 3.5 or 10 mA), in a manner that is visible to a person that may be modifying or extending the circuit in the future.

Starting with the supply requirements, we find in 543.7.1.202 that your supply must comply with one of the following three (1 of 3) methods, if your PE has current levels exceeding 10 mA:

1. Permanently connect the wiring to the equipment (without use of plug and socket-outlet) in accordance with 543.7.1.203 below. You may use a flexible cable to do this.
2. Provide a plug and socket-outlet that complies with BS EN 6309-2, via a flexible chord. Plus one of the following must be met:
   a. The PE for the flexible cable must have a cross-sectional area of:
      i. 2.5 mm^2 for plugs rated at 16 A or less
      ii. 4 mm^2 for plugs rated greater than 16 A
   b. The PE for the flexible cable has a cross-sectional area equal or greater than the line conductor
3. The PE complies with BS 7671 Section 543 *and* has an earth-monitoring device installed to BS 4444 standards that will automatically disconnect the circuit (supply to the equipment) should it detect a line fault.

Now to the circuit requirements we spoke about above (see BS 7671 Section 543.7.1.203). Again, if your protective conductor has amperages (currents) exceeding 10 mA during normal operating conditions, you must have a circuit that complies with one or more of the following (this applies to circuits feeding one or more pieces of equipment):

1. The circuit has a single PE with a cross-sectional area of at least 10 mm^2. This PE must comply with BS 7671 Sections 543.2 and 543.3.
2. The circuit has a single copper PE with a cross-sectional area of at least 4 mm^2. This PE must comply with BS 7671 Sections 543.2 and 543.3, and must be completely enclosed in conduit, flex conduit, and so on, so that it is protected from mechanical damage.

3. The circuit has two PE conductors, or similar or differing materials, each complying with BS 7671 Section 543, AND located in the same metal conduit together.

    Note 1: If the two PE conductors are part of a multicore cable, the total cross-sectional area of every conductor in the cable (including the line and neutral conductors) must be at least 10 mm² to qualify. The PE may be part of the sheath, armour, or wire braid that is incorporated in the assembly of the multicore cable assembly (must comply with 543.2.5).

    Note 2: The ends of the two PE conductors must be terminated independently at all points within the circuit, and may require an accessory with two separate earth terminals. See 543.7.1.204.

4. The circuit has an earth-monitoring device installed to BS 4444 standards that will automatically disconnect the circuit (supply to the equipment) should it detect a line fault.

5. The circuit is connected to a double-wound transformer or equivalent (such as a motor alternator), where the PE for the incoming supply is connected to the exposed conductive parts of the equipment and to the secondary winding of the transformer. This PE conductor must comply with one of the arrangements in 1 through 4 above.

When designing a final circuit with a number of socket-outlets that are likely to run two or more pieces of equipment that will place more the 10 mA on the protective conductor, the circuit must be provided with a high-integrity PE connection complying with 543.7.1, and the following:

1. For final ring circuits that have a PE conductor and a spur (a branch from the ring circuit), the spur must have a high-integrity PE conductor complying with 543.7.1.

2. For radial circuits (any circuit that is *not* a ring) with a single PE conductor:

    a. The PE conductor must be connected as a ring, or

    b. There is a separate PE conductor bonded to the socket-outlet's metal enclosure, conduit, ducting, and so on (i.e., there are two PE conductors provided, one for the socket-outlet's earthing terminal, and a second PE conductor for the metal enclosure, conduit, etc.)

    c. If the distribution board is supplying two or more radial circuits with socket-outlets that are expected to have more than 10 mA of current on the associated PE conductors, and the circuits have similar short-circuit and overcurrent protection devices, and the PE conductors have the same cross-sectional area, then you may do the following: the second PE conductor that is required from above, come from the adjacent circuits PE conductor (instead of running two PE wires, you can share the PE conductors from two similar circuits).

3. Other circuits that comply with 543.7.1.

PROTECTIVE BONDING CONDUCTORS

In Section 544 of BS 7671, we find information regarding protective bonding conductors. Protective bonding conductors are used for bonding exposed (accessible) conductive metal parts together, in order to provide protection for persons from electric shock hazards. Chapter 4 of BS 7671 discusses the requirement to have these conductors, while this section discusses the physical and installation requirements of the protective bonding conductors. In particular, BS 7671 Section 411.3.1.2, Protective Equipotential Bonding, and 411.4 to 411.6 should be reviewed.

MAIN PROTECTIVE BONDING CONDUCTORS

Main protective bonding jumpers are any protective conductor that is routed from the MET to any other extraneous conductive part such as building steel, water pipe, gas line, electronic communication system, and so on.

Section 544.1 of BS 7671 has rules for protective bonding conductors for installations that are PME, and those that are not. This section also exempts highway power supplies and street furniture from its requirements. PME is a typical type of system used by the electrical supplier, where in a single transformer feed multiple services using a PEN conductor (TN-C-S), and may be earthed at each service plus the transformer, hence the multiple earthed arrangement.

Protective multiple earthing (PME)   An earthing arrangement, found in TN-C-S systems, in which the supply neutral conductor is used to connect the earthing conductor of an installation with Earth, in accordance with the Electricity Safety, Quality, and Continuity Regulations (2002).

For installations that are *not* a PME arrangement, according to 544.1.1 all of your main protective bonding conductors must have a cross-sectional area that is at least half the required cross-sectional area of the installations earthing conductor (see BS 7671 Section 542.3), and not less than 6 mm². If your protective bonding conductor is to be made of copper (and it really should be), then this conductor need not be larger than 25 mm² regardless of the earthing conductors size.

For installations that are PME arrangements (the majority of installations will be PME), your main protective bonding conductors are to be sized according to Table 54.8. Note that the selection is based on the size of the incoming (supply) neutral conductor.

Table 54.8   Minimum Cross-Sectional Area of the Main Protective Bonding Conductor in Relation to the Neutral of the Supply

Copper Equivelant Cross-Sectional Area of the Supply Neutral Conductor	Minimum Copper Equivelant* Cross-Sectional Area of the Main Protective Bonding Conductor
35 mm² or less	10 mm²
over 35 mm² up to 50 mm²	16 mm²
over 50 mm² up to 95 mm²	25 mm²
over 95 mm² up to 150 mm²	35 mm²
over 150 mm²	50 mm²

Note: Local distributor's network conditions may require a larger conductor
*The minimum copper equivalent cross-sectional area is given by a copper bonding conductor of the tabulated cross-sectional area or a bonding conductor or another metal affording equivelant conductance

We have some very specific locations where the main protective bonding conductor is to be installed on our water, gas, and other services (telephone, data, etc.). Namely, the connection must be made as near as practical to the entry point of the service into the premises. Furthermore, the connection must be made to the consumer's hard metal pipework past any insulating joints or connections made by the supplier, and before any branch work of the system. Preferably, the connection will be made within 600 mm (2 ft) of the meter outlet union or within 600 mm of the entry to the building (where the meter is external).

SUPPLEMENTARY BONDING CONDUCTORS

Supplementary bonding conductors are those conductors that are added in addition to the main protective bonding conductors to ensure fault protection is maintained throughout a given system. An example would be an extra bonding conductor installed around a plastic water filter located below a kitchen sink (located entirely within the consumer's premises). Maintaining the electrical continuity of the metal faucets people will be touching in a wet sink area that is located directly adjacent to electrically powered equipment, is critical to human safety.

We find in 544.2 that the basic rule is that the supplementary bonding conductors are to have a cross-sectional area that is at least half that of the smaller of the related PE conductors for the installation, but must additionally meet the following:

- If the supplemental bonding conductor is protected from mechanical damage (such as routed in conduit), then its minimum cross-sectional area must be at least 2.5 mm².
- If the supplemental bonding conductor is *not* protected from mechanical damage (such as routed exposed), then its minimum cross-sectional area must be at least 4 mm².

For fixed appliances (such as a stove) supplied with a short length of flexible cable, the PE conductor installed within the flexible cable will suffice for meeting the supplementary bonding conductor requirement found in 544.2, as stated in 544.2.5.

Chapter Seven

PART 6: INSPECTION AND TESTING

Part 6 of BS 7671 deals with the inspection and testing of electrical installations. In regards to our topic of earthing, there are really only three tests: earth electrode resistance, earth fault loop impedance (Z_s), and external earth fault loop impedance (Z_e).

Before we proceed into testing, it is important to take a look back to Part 4 of BS 7671 and review the fundamental safety requirements of the code.

The basic fundamentals for electric safety are quite simple:

1. Make hazardous-live-parts inaccessible to persons and livestock
2. Earth all exposed conductive parts and bond them together so that there are no differences in potential and to protect against electromagnetic disturbances
3. Provide protection devices to automatically disconnect power in the event of a fault
4. Provide devices to automatically remove overvoltages

The best way to protect persons and livestock from electric shock is to earth all exposed conductive parts and bond them together to ensure that there are no differences in potential.

The tests found in Part 6 of the Code are intended to help ensure that the requirements found in Part 4 of the Code are met.

In regards to the three tests themselves, they can be broken down as follows:

- **Earth electrode resistance** determines how effectively earthed a given electrode is in relation to a remote point of earth, in ohms.
- **Earth fault loop impedance (Z_s)** is a test commonly conducted on a branch circuit downstream from the distribution board, such as at a socket-outlet. It is used to measure the effectiveness of the *furthest-point* of a circuit's protective conductor (PE) back to the distribution board.
- **External earth fault loop impedance (Z_e)** is a test commonly conducted on the facilities main incoming electrical service. It is designed to check the effectiveness of the PE or PEN conductor upstream from the facilities main electrical distribution board to the suppliers (electric utility) transformer. This test is often referred to as a test of the effectiveness of the distributor's earth.

Please note that some of the tests discussed below are required to be conducted on a live circuit, which can be very dangerous if not conducted by skilled or competent persons specifically trained to conduct these tests. There are even some legal requirements under HSR25 EWR Regulation 14 for personnel conducting these tests to have proper training for working on live conductors, having the correct and proper personal protective gear, having electrical lock-out tang-out equipment, and other injury prevention and treatment precautions available while conducting these tests.

Conducting tests on electrically live circuits is a dangerous activity that falls outside of the scope of this book. Below we will discuss the basics of the testing, however, specific methodology on test methods and proper electrical safety techniques will not be discussed. If you are conducting such tests, make sure that you are properly trained and certified to meet all of the legal requirements needed to work on electrically live circuits.

EARTH ELECTRODE RESISTANCE

The earth electrode resistance, or sometimes called the resistance-to-earth or resistance-to-ground, is a measurement of earth electrodes resistance in ohms compared to remote earth. There are a number of electrical test methods that can be used to measure the earth electrode resistance. See the discussion on resistance-to-earth measurements in this book, or in BS 7430:2011+A1:2015 Section 10.3.

Note: The driving force behind the earth electrode resistance is the soil resistivity. You do not get to choose what type of soil/earth your installation is sitting on. An earth electrode just provides a metallic connection across a given length of resistive soils, what the resistivity of those soils are is not in your control. The point is that if you are trying to achieve a desired earth resistance for your electrode installation, say 5 Ω, then it must be engineered in advance from soil resistivity data that was collected on site. Trying to achieve a given earth electrode resistance during the construction phase by randomly driving rod after rod, is simply bad engineering.

In Section 612.7, we find a simple requirement to measure the earth-electrode-resistance (R_A) of all earth electrodes that are required to be installed to meet the regulation for our system. There is no listed minimum resistance for your electrode in either BS 7671 or in BS 7430. This tests does make up part of the earth fault loop impedance test, which does have a maximum of 200 Ω under BS 7671 Table 41.5 Note 2. Therefore, for a TT system the earth electrode would need to be something significantly lower than 200 Ω in order to qualify.

Note: Under the U.S. electrical code, a single earth electrode is required to measure at 25 Ω or less.

This means that if you have a TT system, you must measure the earth electrode resistance (R_A) of the installed earth electrode, as this electrode is mandatory.

However, if you have a TN-C-S system and have optionally installed an earth electrode (for extra safety) that was not mandatory, you need not measure the earth electrode resistance.

For further information regarding how to properly conduct an earth electrode resistance test/measurement, please see the discussion on resistance-to-earth measurements in Chapter 9 of this book, or in BS 7430:2011+A1:2015 Section 10.3.

EARTH LOOP IMPEDANCE TEST—GENERAL

The earth fault loop impedance is the path followed by fault current when a low impedance fault occurs between the phase conductor and earth, that is, "earth fault loop." Fault current is driven round the loop by the supply voltage.

We have several sources that govern the earth loop impedance tests (both the Z_s and the Z_e), namely:

- BS 7671:2008+A3:2015 Section 612.9
- BS 7671 Appendix 14
- *On-Site Guide* BS 7671:2008+A3:2015 Section 10.3.6
- IET *Guidance Note 3: Inspection & Testing*

Note: BS 7430:2011+A1:2015 does not discuss either of the earth loop impedance tests.

In BS 7671 Section 612.9, we find the requirement to conduct the "relevant" earth loop impedance tests on circuits where protective measures are used. This simple one-sentence line of instruction drives the entire earth loop impedance test requirement, both the Z_s and the Z_e. We must then turn our attention to the *On-Site Guide* BS 7671:2008+A3:2015 Section 10.3.6.

EARTH LOOP IMPEDANCE TEST (Z_s)

Earth fault loop impedance (Z_s) is a test that is required to be conducted on every circuit installed within a facility (downstream from the distribution board). This test must be conducted at the furthest-point of each branch circuit to ensure that the PE conductor is working effectively.

We find in the *On-Site Guide* BS 7671:2008+A3:2015 Section 10.3.6 that we have three methods that can be used to measure the earth fault loop impedance (Z_s):

1. Conduct a direct measurement of the Z_s using a specially designed electrical meter. This is the most common method as it is reliable, quick, and easy to do.
2. Conduct a series of tests, including measuring the Z_e, and calculate the result.
3. Conduct a series of tests, using the declared Z_e from the distributor, and calculate the result.

So what are these series of other tests discussed in 2 and 3 above? The first one is the Z_e that we will be discussing in the section below. The other tests, and there are two of them, are found in the *On-Site Guide* Sections 10.3.1 and 10.3.2. One is for regular circuits, and the other is for ring circuits.

- 10.3.1 – Continuity of circuit protective conductors and protective bonding conductors
- 10.3.2 – Continuity of ring final circuit conductors

Both tests provide R_1 and R_2 data that is needed for us to complete the following formula:

$$Z_s = Z_e + (R_1 + R_2)$$

If the electrical distributor has provided us with a declared Z_e, then we should use that number in the equation above. If not, we will need to measure for Z_e as discussed below.

However, it is highly unlikely that you will be using this formula, as in today's electronic environment, this test should be directly measured by a test meter specifically designed for measuring the Z_s.

The Z_s test must be conducted on a live circuit, which can be very dangerous if not conducted by skilled or competent persons specifically trained to conduct these tests. There are even some legal requirements under HSR25 EWR Regulation 14 for personnel conducting these tests to have proper training for working on live conductors, having the correct and proper personal protective gear, having electrical lock-out tang-out equipment, and other injury prevention and treatment precautions available while conducting these tests.

EXTERNAL EARTH LOOP IMPEDANCE TEST (Z_E)

Earth fault loop impedance is the path followed by fault current when a low impedance fault occurs between the phase conductor and earth, i.e., "earth fault loop." Fault current is driven round the loop by the supply voltage. In the case of the Z_e, we are measuring the external earth loop impedance, which is a measurement of the fault current loop back to the distributor's transformer, typically located outside of our facility (and may be a PME system).

The Z_e test requires that the main electrical power be turned off for the entire facility at the main distribution board. However, the incoming distributors (electrical utility) power lines will still be energized from the main transformer. Which means that this test must be conducted on a live circuit with the circuit overload protection controlled by the distributor at a remote location, which can be extremely dangerous if not conducted by skilled or competent persons specifically trained to conduct these tests. There are even some legal requirements under HSR25 EWR Regulation 14 for personnel conducting these tests to have proper training for working on live conductors, having the correct and proper personal protective gear, having electrical lock-out tang-out equipment, and other injury prevention and treatment precautions available while conducting these tests.

To conduct this test, we really need a test meter that is specifically designed for measuring the Z_e. This test requires you to enter the main distribution box and lock-out tag-out the main incoming power lines for the entire facility, and then remove the main earthing conductor from the earth terminal. The test itself measures from the main earthing conductor to the line wires, to provide the measured Z_e.

On occasion, the electrical distributor will declare what the Z_e for a given location.

Again, since these tests are conducted on electrically live components, the specifics on methodology and safety fall outside of the scope of this book. Proper training is mandatory under HSR25 EWR Regulation 14.

CONSIDERATION FOR THE INCREASE IN RESISTANCE OF CONDUCTORS WITH THE INCREASE IN TEMPERATURE

In Appendix 14 of BS 7671, we find the formula needed for calculating the increase in resistance of our Z_s, given an increase in temperature. In other words, if we are measuring the Z_s during winter conditions, it will not be the same as if we measured the Z_s in summer conditions, and we must accommodate for those changes.

This requirement is for TN and TT systems and the measurements must be made in compliance with 411.4 and 411.5. The following formula must be completed:

$$Z_s(m) = 0.8 * \frac{U_0 * C_{min}}{I_a}$$

where:

$Z_s(m)$ = measured earth fault current loop in ohms, from the most distant point of the relevant circuit from the origin of the installation

U_0 = nominal AC rms line voltage to earth in volts

I_a = current in amperes (A) that will cause the protective device to trip according to BS 7671 Table 41.1 or within 5 seconds per 411.3.2.3(A)

C_{min} = minimum voltage factor to take account of voltage variations depending on time and place, changing of transformer taps, and other considerations
Note: For a low voltage supply the value is 0.95, in accordance with Electricity Safety, Quality, and Continuity Regulations (2002).

If the result of the calculation above gives a value for the earth fault loop impedance that is greater than 0.8 U_0/I_a, a detailed assessment of the earth fault loop impedance is required under 411.5 or 411.5.4. The following procedure is to be used:

1. Measure the line to PE conductor loop at the origin of the installation.
2. Measure the distribution circuits line and PE conductors individual resistances.
3. Measure the final circuits line and PE conductors individual resistances.
4. Increase the measured resistances from b and c above, based on the increased in resistance due to expected conductor temperature, both for environmental temperature increases and for the anticipated current load on the conductor temperature increase.
5. Add the values of a and d above, to get a new realistic estimated Z_s.
6. Other methods may be acceptable.

APPENDIX B TABLES

The following tables are from Appendix B of the *On-Site Guide* and provide the maximum permissible measured earth fault loop impedances (Z_s). All figures in these tables are based on an ambient temperature of 10°C and a nominal voltage to earth (U_0) of 230 V. If measurements are made at temperatures greater than 10°C, a correction factor found in Table B8 must be multiplied to the measured Z_s in order to get a result.

Tables B1 to B5 can be used to find the maximum earth fault loop impedances for the following cables, if the cable loading is such that the maximum operating temperature is 70°C, and if they meet a few additional requirements:

- Cables from BS 6004, Tables 3, 4, and 5
- Cables from BS 7211, Tables 3, 4, and 5
- Cables from BS EN 5025-3-41, Tables B1 and B2
- Other thermoplastic (PVC) cables
- Other thermosetting/low smoke halogen free (LSHF) cables

Those additional requirements are:

1. The PE conductor is made of copper
2. The PE conductor has a cross-sectional area between 1 and 16 mm^2
3. The overcurrent protective device is rated to one of Tables B1 to B5 listed in this chapter

Additional information can be found in BS 7671 Sections 411.3.2.2–3, 411.4.6–8, 543.1.3, and 612.9 and the IET Commentary on the Wiring Regulations.

Note: The values presented in the tables of Appendix B are based on a 10°C conductor temperature, while Tables 41.2, 41.3, and 41.4 in BS 7671 Chapter 41 are based on a higher conductor temperature.

Table B1   Fuses

Semi-enclosed fuses. Maximum measured earth fault loop impedance (in ohms) at ambient temperature where the overcurrent protective device is a semi-enclosed fuse to BS 3036

0.4 Second Disconnection (Final Circuits Not Exceeding 32A in TN Systems)

Protective Conductor in mm^2	Fuse Rating			
	5 A	15 A	20 A	30 A
1.0	7.3	1.9	1.3	NP
≥ 1.5	7.3	1.9	1.3	0.83

5 Second Disconnection (Final Circuits Exceeding 32A and Distrubtion Circuits in TN Systems)

Protective Conductor in mm^2	Fuse Rating			
	20 A	30 A	45 A	60 A
1.0	2.3	NP	NP	NP
1.5	2.91	1.6	NP	NP
2.5	2.91	2.0	1.0	NP
4.0	2.91	2.0	1.2	0.85
≥ 6	2.91	2.0	1.2	0.85

NP = Combination of fuse and PE is not permitted

Table B2   Fuses

BS 88-2.2 and BS 88-6 fuses. Maximum measured earth fault loop impedance (in ohms) at ambient temperature where the overcurrent protective device is a semi-enclosed fuse to BS 88-2.2 or BS 88-6

0.4 Second Disconnection (Final Circuits Not Exceeding 32A in TN Systems)

Protective Conductor in mm²	Fuse Rating					
	6 A	10 A	16 A	20 A	25 A	32 A
1.0	6.47	3.90	2.06	1.34	1.09	0.62
1.5	6.47	3.90	2.06	1.34	1.09	0.79
≥ 2.5	6.47	3.90	2.06	1.34	1.09	0.79

5 Second Disconnection (Final Circuits Exceeding 32A and Distrubtion Circuits in TN Systems)

Protective Conductor in mm²	Fuse Rating							
	20 A	25 A	32 A	40 A	50 A	63 A	80 A	100 A
1.0	1.46	1.17	0.62	NP	NP	NP	NP	NP
1.5	2.03	1.40	1.00	0.60	NP	NP	NP	NP
2.5	2.21	1.75	1.40	0.81	0.70	0.34	NP	NP
4.0	2.21	1.75	1.40	1.03	0.76	0.49	0.24	NP
6.0	2.21	1.75	1.40	1.03	0.79	0.62	0.34	0.19
10.0	2.21	1.75	1.40	1.03	0.79	0.62	0.44	0.29
16.0	2.21	1.75	1.40	1.03	0.79	0.62	0.44	0.32

NP = Combination of fuse and PE is not permitted

Table B3   Fuses

BS 88-2 fuses. Maximum measured earth fault loop impedance (in ohms) at ambient temperature where the overcurrent protective device is a semi-enclosed fuse to BS 88-2

0.4 Second Disconnection (Final Circuits Not Exceeding 32A in TN Systems)

Protective Conductor in mm²	FuseRating							
	2 A	4 A	6 A	10 A	16 A	20 A	25 A	32 A
1.0	26.5	12.5	6.2	3.7	1.9	1.3	1.0	0.6
1.5	26.5	12.5	6.2	3.7	1.9	1.3	1.0	0.8
≥ 2.5	26.5	12.5	6.2	3.7	1.9	1.3	1.0	0.8

(Continued)

Table B3   Fuses *(Continued)*

5 Second Disconnection (Final Circuits Exceeding 32A and Distrubtion Circuits in TN Systems)

Protective Conductor in mm²	Fuse Rating							
	20 A	25 A	32 A	40 A	50 A	63 A	80 A	100 A
1.0	1.46	1.03	0.63	0.55	NP	NP	NP	NP
1.5	2.13	1.20	0.87	0.83	NP	NP	NP	NP
2.5	2.24	1.70	1.40	1.00	0.50	0.30	NP	NP
4.0	2.24	1.70	1.40	1.00	0.76	0.49	0.22	0.12
6.0	2.24	1.70	1.40	1.00	0.79	0.62	0.30	0.19
10.0	2.24	1.70	1.40	1.00	0.79	0.62	0.44	0.32
16.0	2.24	1.70	1.40	1.00	0.79	0.62	0.44	0.34

NP = Combination of fuse and PE is not permitted

Table B4   Fuses

BS 88-3 fuses. Maximum measured earth fault loop impedance (in ohms) at ambient temperature where the overcurrent protective device is a semi-enclosed fuse to BS 88-3

0.4 Second Disconnection (Final Circuits Not Exceeding 32A in TN Systems)

Protective Conductor in mm²	Fuse Rating			
	5A	16 A	20 A	32 A
1.0	7.90	1.84	1.55	0.60
1.5 to 1.6	7.90	1.84	1.55	0.73

5 Second Disconnection (Final Circuits Exceeding 32A and Distrubtion Circuits in TN Systems)

Protective Conductor in mm²	Fuse Rating					
	20 A	32 A	45 A	63 A	80 A	100 A
1.0	2.13	0.59	NP	NP	NP	NP
1.5	2.57	0.76	NP	NP	NP	NP
2.5	2.57	1.13	0.55	0.24	NP	NP
4.0	2.57	1.25	0.76	0.32	0.19	NP
6.0	2.57	1.25	0.76	0.51	0.29	0.16
10.0	2.57	1.25	0.76	0.55	0.40	0.26
16.0	2.57	1.25	0.76	0.55	0.40	0.30

NP = Combination of fuse and PE is not permitted

Table B5   Fuses

BS 1361 fuses. Maximum measured earth fault loop impedance (in ohms) at ambient temperature where the overcurrent protective device is a semi-enclosed fuse to BS 1361

0.4 Second Disconnection (Final Circuits Not Exceeding 32A in TN Systems)

Protective Conductor in mm²	Fuse Rating			
	5 A	16 A	20 A	32 A
1.0	8.00	2.50	1.29	0.70
1.5 to 1.6	8.00	2.50	1.29	0.86

5 Second Disconnection (Final Circuits Exceeding 32A and Distrubtion Circuits in TN Systems)

Protective Conductor in mm²	Fuse Rating					
	20 A	32 A	45 A	60 A	80 A	100 A
1.0	1.46	0.70	NP	NP	NP	NP
1.5	1.98	0.97	0.30	NP	NP	NP
2.5	2.13	1.40	0.49	0.20	NP	NP
4.0	2.13	1.40	0.67	0.35	0.25	NP
6.0	2.13	1.40	0.73	0.47	0.20	0.12
10.0	2.13	1.40	0.73	0.53	0.38	0.20
16.0	2.13	1.40	0.73	0.53	0.38	0.28

NP = Combination of fuse and PE is not permitted

CIRCUIT BREAKERS

Table B6   Circuit Breakers

Circuit-breakers. Maximum measured earth fault loop impedance (in ohms) at ambient temperature where the overcurrent protective device is a circuit-breaker to BS 3871 or BS EN 60898 or RCBO to BS EN 61009

0.1 to 5 Second Disconnection Times

Circuit-Breaker Type	Circuit-Breaker Ratind in Amperes														
	3	5	6	10	15	16	20	25	30	32	40	45	50	63	100
1	14.56	8.74	7.28	4.40	2.93	2.76	2.20	1.76	1.47	1.38	1.10	0.98	0.88	0.70	0.44
2	8.40	5.00	4.20	2.50	1.67	1.58	1.25	1.00	0.83	0.79	0.63	0.56	0.50	0.40	0.25
B	11.65	7.00	5.87	3.50	2.30	2.20	1.75	1.40	1.17	1.10	0.88	0.78	0.70	0.56	0.35
3&C	5.82	3.49	2.91	1.75	1.16	1.09	0.87	0.70	0.58	0.55	0.44	0.38	0.35	0.27	0.17

(Continued)

Table B6 Circuit Breakers *(Continued)*

Circuit-breakers. Maximum measured earth fault loop impedance (in ohms) at ambient temperature where the overcurrent protective device is a circuit-breaker to BS EN 60898 type D or RCBO to BS EN 61009 type D										
Circuit-Breaker	Circuit-Breaker Ratind in Amperes									
	6	10	16	20	25	32	40	50	63	100
D 0.4 Seconds	1.46	0.87	0.55	0.44	0.35	0.28				
D 5 Seconds	2.91	1.75	1.09	0.87	0.70	0.55	0.44	0.35	0.28	0.17

Note: BS 7671 Regulation 434.5.2 requires that the PE conductor csa (cross-section area) meet the requirments of BSEN 60898-1, 2 or BSEN 61009-1, or the minimum quoted by the manufacturer.

MINIMUM PROTECTIVE CONDUCTOR (PE) SIZE

Table B7

Minimum Protective Conductor Size in mm^2			
Energy Limiting Class 3 Device Rating	Fault Level in kA	Protective Conductor Cross-Sectional Area (csa) in mm^2	
		Type B	Type C
Up to and including 16 A	≤ 3	1.0	1.5
Up to and including 16 A	≤ 6	2.5	2.5
Over 16 up to and including 32 A	≤ 3	1.5	1.5
Over 16 up to and including 32 A	≤ 6	2.5	2.5
40 A	≤ 3	1.5	1.5
40 A	≤ 6	2.5	2.5

Note: Consult manufacturers data for higher fault levels or other device types and ratings. See BS 7671 Regulation 434.5.2 and IET Commentary on the IET Wiring Regulations for further information.

TEMPERATURE CORRECTION FACTORS

The correction factor is to be simply multiplied to your Z_s in order to get a temperature corrected Z_s.

Table B8

Ambient Temperature Correction Factors	
Ambient Temperature in °C	Correction Factor from 10°C
0	0.96
5	0.98
10	1.00
20	1.04
25	1.06
30	1.08

Note: This table is different form Table I2 because this table corrects from 10°C and Tabel I2 corrects from 20°C.

Chapter Eight

ELECTRODE CALCULATIONS

Note: This chapter was co-written by Svetlana Knyazeva-Johnson.

INTRODUCTION

Grounding (earthing) is the intentional connection of your electrical system to the earth (soil). The quality of this connection has been known since the late 1800s to be fundamentally important for ensuring the safe and effective operation of an electrical system, during both normal operating conditions and abnormal fault conditions. This electrical connection to earth can be measured in the form of resistance and/or impedance by referencing the grounding system in question to a remote source, and injecting a known electrical signal into the grounding system and propagating it across the earth. A proper measurement of this type will give us a resistance/impedance to ground of our grounding/earthing electrode system.

It is important to understand that the resistivity of soil changes dramatically from one location of the globe to another, ranging from very low resistivities of around 10 Ωm to as high as a million Ωm, with typical resistivities ranging between 20 and 10,000 Ωm. As such, engineers are concerned about how effectively connected their electrical systems are to earth, and often have specifications that require their grounding systems to meet a specific resistance to ground (RTG). Common specifications include the requirements to measure the earth fault loop impedance in BS 7671 Appendix 14 (especially for TT systems), the U.S. National Electrical Codes' 25-Ω RTG requirement, and the telecommunication industries' 5-Ω RTG requirement. These specifications require engineers to design grounding electrode systems well in advance of the installation, to ensure compliance.

Grounding electrode systems with high resistances to earth are susceptible to a number of harmful electrical issues, including increased electrical noise such as harmonics and transients, increased step and touch voltage hazards, and most importantly the system feeders from electrical utility companies rely on the earth as the fault current return path. During fault conditions, high-resistance grounding systems can force objectionable currents onto other connected systems, such as telephone lines, water lines, CATV systems, gas lines, and so on, and decrease the ability of the utility to detect and deal with electrical fault situations. The most common grounding/earthing calculations are designed to discover the overall resistance and/or impedance of their grounding electrode system prior to installation, to prevent these undesirable electrical issues from occurring. Of note is that the Canadian Electrical Code (CEC) uses a 5000-V ground potential rise (GPR) limit in contrast to the 25-Ω RTG used by the National Electrical Code (NEC).

The primary goal of most of the calculations in this section is to estimate in advance what the resistance/impedance to ground of the electrode system or grounding grid will be, so that once the electrical system is conducted, it meets the required minimal standards.

The calculations presented in this section are simple hand calculations, primarily based on theoretical uniform soil conditions with direct current (DC) electrical systems and are intended for rough estimations only. Caution should be used when conducting any calculation involving human safety, as BS EIC 60050-195, 195-05-11, modified, and IEV 195-05-12 (in the U.S. Federal law 29 CFR 1910.269) have strict requirements for ensuring the safety of personnel working in hazardous electrical environments, and the simple calculations presented below are not adequate for ensuring the safety of humans. In today's computer age, it is my opinion that it is unethical to conduct hand calculations versus using computer-modelling software.

When modelling an electrical system, the accuracy of the model is generally improved with each additional soil layer that is added. In my field experience, computer models that utilize three to five soil layers when making its calculations provide the most accurate results. There are a number of computer software programs available in the market place that are capable of analyzing grounding systems in not only theoretical uniform soil, but also in two-layer soil resistivity conditions. However, only a few programs are capable of analyzing soil conditions with three or more soil layers, and fewer still that can properly analyze touch voltages. With so many engineering factors that must be taken into account, careful selection must be made when selecting a grounding software program.

With that being said, this section will discuss the basic formulas used in calculating grounding/earthing factors. While there is just too much information on this subject to simply cover in a single section, the following formulas have been selected to provide you an overall understanding of the physics behind ground system engineering.

There is a great deal of additional information regarding proper test methods, grounding electrodes, bonding, and proper wiring techniques that should be well understood in addition to understanding the formulas presented here. Please see *McGraw-Hill's Standard Handbook of Electrical Engineers* or *McGraw-Hill's National Electrical Code 2014 Earthing & Grounding Handbook* for more information.

BASIC CALCULATION PROCESS

The basic process of analyzing a grounding system starts by determining what type of electrical system the grounding system you are designing will be used for DC or alternating current (AC). If you are designing a grounding electrode system for a DC electrical system, then you only need to conduct resistance equations. If, however, your electrical system is AC, you will need to use the impedance equations that are more intensive.

Note: Unless you are modelling lightning or other high-frequency noise, 50-/60-Hz systems have a frequency that is low enough that you can generally model them using the DC equations.

The next step is to calculate the resistance/impedance of your proposed grounding electrode system. This can take a number of calculations based on your design. You will find in the pages below multiple formulas for assisting in this process. Unfortunately, most of the formulas are for DC electrical systems, as the impedance calculations are simply too complex to conduct by hand. The only impedance equation presented in this section is for a single ground rod and is shown as an example of how complex the formulas become once ACs are added to the equation.

In order to calculate the resistance/impedance of your proposed grounding system, you will need a soil resistivity model of the proposed installation site. A soil resistivity model provides calculated soil layer depths with resistivities, and is derived from the raw data gathered at the site by conducting a soil resistivity test, such as a Wenner four-point test. Once you have a valid soil resistivity model, you will know what type of soils you are dealing with, uniform (single layer), two layers, three layers, or more. When modelling an electrical system, the accuracy of the model is generally improved with each additional soil layer that is added. In my field experience, computer models that utilize three to five soil layers when making its calculations provide the most accurate results.

Note that there is a difference between a calculated soil resistivity model, and apparent resistivity as calculated using a simple formula. You need a soil model, not apparent resistivity. Please see the section on soil resistivity for more information.

Now that you know if you are calculating impedance or resistance, the soil resistivities, and how many soil layers you have, you can now select the equations you must complete in order to calculate the resistance/impedance of your proposed grounding system (as stated earlier, if your electrical system is AC, or you have two or more soil layers, you will really need a computer). In the pages below you will find a series of equations to help you figure out the overall resistance/impedance to ground of your grounding electrode system.

Once you have determined the resistance/impedance to ground, you can then calculate the GPR of the electrode system at your site. These GPR equations calculate what the voltage will be on your grounding system should an electrical fault occur. As such, the GPR equations require actual electrical-fault data to be combined with the resistance-/impedance-to-ground data for your site. Some countries, such as Canada, actually have a 5000-V GPR limit for grounding system installations, and calculations must be provided. In the GPR section of this chapter, you will find basic formulas for conducting these equations.

If your electrical system uses high-voltage (1000 V or higher), delta-type transformers, on-site separately derived power generators, or has other special human safety concerns, you will need to conduct step and touch voltage equations to determine if the site needs additional grounding/earthing to make it safe for personnel. These equations will require resistance-/impedance-to-ground and the GPR results from above along with a number of other factors.

If your site has a human safety issue due to step and touch voltages, Federal Laws such as 29 CFR 1910.269 may require you to contact an engineering firm specializing in these issues.

BACKGROUND

There are a number of important grounding concepts that must be understood in order to properly analyze your grounding system. This section will only introduce you to a small part of these important concepts, please see *McGraw-Hill's Standard Handbook of Electrical Engineers* or *McGraw-Hill's National Electrical Code 2014 Earthing & Grounding Handbook* for more information.

Zone (or Sphere) of Influence. An important concept as to how efficiently grounding electrodes discharge electrical currents into the earth is called "the zone of influence," which is sometimes referred to as the "sphere of influence" (see Fig. 8.1). The zone of influence is the volume of soil throughout which the electrical potential rises to more than a small percentage of the potential rise of the ground electrode when that electrode discharges current into the soil. The greater the volume compared with the volume of the electrode, the more efficient the electrode. Elongated electrodes, such as ground rods, are the most efficient. The surface area of the electrode determines the ampacity of the device but does not affect the zone of influence. Case in point, the greater the surface area, the greater the contact with the soil and the more electrical energy that can be discharged per unit of time.

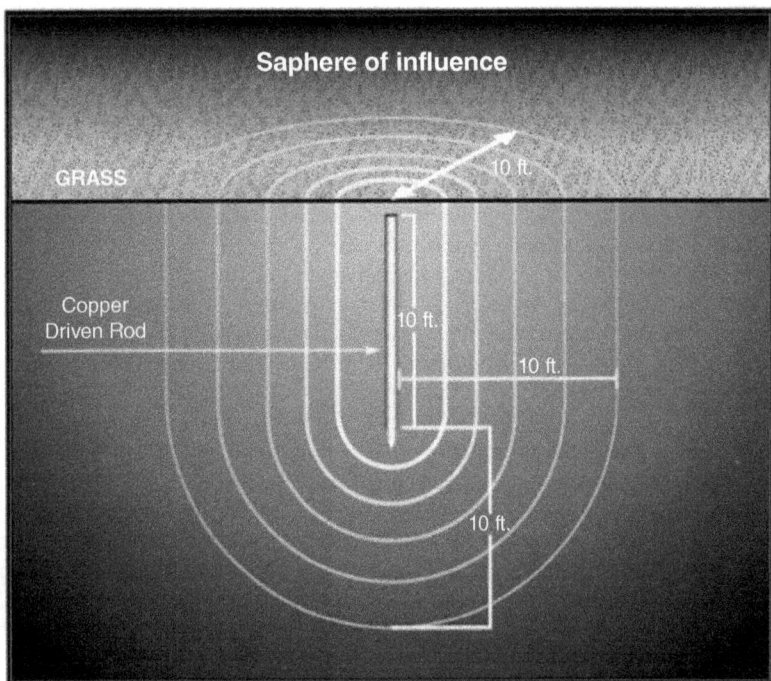

Figure 8.1 Sphere of influence.

The formula for calculating the volume of soil utilized by a given vertical electrode is:

$$V = \frac{5\pi L^3}{3}$$

where V = volume of soil, L = length of electrode in feet or metres, and π = 3.142. A simpler version of the above formula can be generated by rounding π (pi) down to 3, and cross-cancelling to get the formula:

$$V = 5L^3$$

where V = volume of soil and L = length of electrode in feet or metres.

Thus, a single 10-ft driven rod will utilize 5000 ft³ of soil, whereas a single 8-ft rod will utilize about half the soil at 2560 ft³. Going from an 8- to 10-ft ground rod can provide a significant reduction in the RTG, as the sphere of influence will be nearly doubled, given that soil resistivity does not increase with depth. This is why you want to space your 10-ft ground rods at least 20 ft apart, so that you do not have overlapping spheres of influence.

These two rules of thumb, deeper ground rods are better and spacing the ground rods at least 2 × their length apart, are valuable guidelines for designing your grounding system.

Soil Resistivity Testing. Soil resistivity testing is the process of measuring a volume of soil to determine the conductivity of the soil. The resulting soil resistivity is expressed in ohm-metres or ohm-centimetres.

Soil resistivity testing is the single most critical factor in electrical grounding design. This is true when discussing simple electrical design, dedicated low-resistance grounding systems, or the far more complex issues involved in GPR studies. Good soil models are the basis of all grounding designs, and they are developed from accurate soil resistivity testing.

The Wenner four-point (or four-pin) method has long been considered the single most accurate method for obtaining valid soil resistivity data. IEEE Std. 81-1983 and ASTM D 6431-99 are excellent sources of information for this test method, and most importantly requires DC test metres that utilize an induced polarization technique capable of 50 W or more.

The Wenner four-point soil resistivity test method utilizes four probes spaced at equidistant intervals, called the "A spacing." The maximum "A spacing" is typically equal to or larger than maximum diagonal distance of the proposed ground grid. Numerous smaller "A spacings" are then taken to obtain a valid soil model of the site. The variation of these "A spacing" intervals is critical to obtaining a proper soil model; please see the above-mentioned standards and/or *McGraw-Hill's Standard Handbook of Electrical Engineers* or *McGraw-Hill's National Electrical Code 2014 Earthing & Grounding Handbook* for more information.

Apparent Resistivity Equation. Numbers that are generated from the apparent resistivity equation are often confused with the numbers that come from a soil model. It is very important to understand the difference and to know that the apparent resistivity equation will *not* provide the soil resistivity numbers you want for accurate calculations. Only a computer-generated soil model will provide the accurate soil resistivity numbers that you will want to use when modelling your electrical system.

Apparent resistivity is defined by the U.S. Environmental Protection Agency as the resistivity of a (fictitious) electrically homogeneous and isotropic half space that would yield the measured relationship between the applied current and the potential difference for a particular arrangement and spacing of electrodes. It is considered fictitious because homogenous (uniform or single-layer soil) does not exist in practice.

Apparent resistivity provides a single soil resistivity number for a given probe spacing, but does not provide information regarding the number of soil layers or their depths.

For example, let us say you have the measured resistances (R) taken from field measurements for spacings of 5 in. and 10 in. (see Fig. 8.2). Using the apparent resistivity equation, you get an apparent resistivity for 5 in. of 50 Ωm, and for the 10 in. spacing you get an apparent resistivity of 75 Ωm. If you think of the spacings of the probes as a function of depth, you would have the average (apparent) resistivity from the surface of the earth (0 in.) to the depth of 5 in. spacing of 50 Ωm (from 0 to 5 in.), and the 0–10 in. apparent resistivity of 75 Ωm.

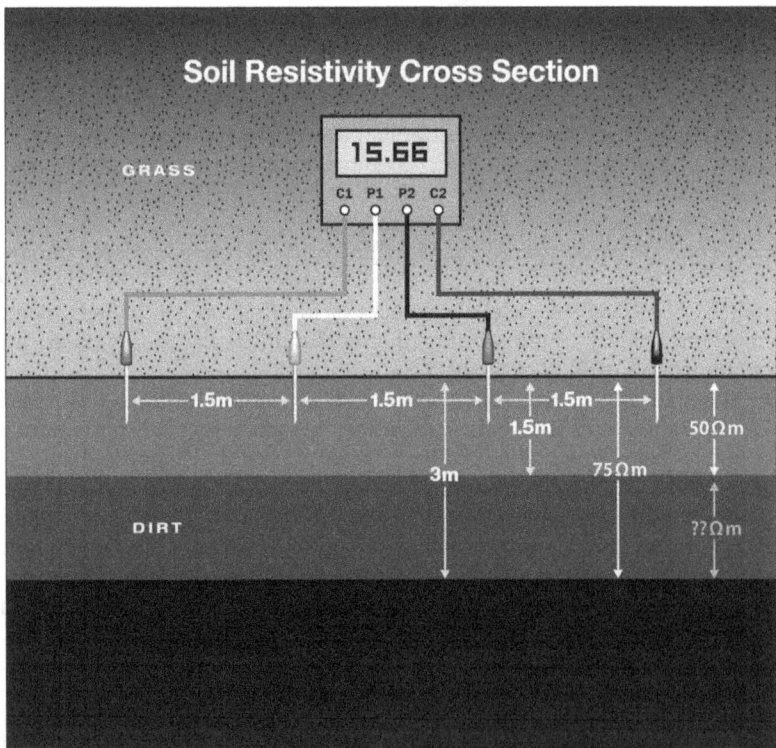

Figure 8.2 Soil resistivity spacings versus depth.

In other words, the 10 in. reading is a combination of the 5 in. data you already have, plus the additional 5 in. of soil below that.

So the question remains, what is the soil resistivity between 5 and 10 in.? The easy answer: the soil between 5 and 10 in. must be 100 Ωm. If the apparent resistivity from 0 to 5 in. is 50 Ωm, and the resistivity from 5 to 10 in. is 100 Ωm, then the apparent resistivity of both the 5 in. of 50-Ωm soil and the 5 in. of 100-Ωm soil, would give you an average of 75 Ωm.

Again, apparent resistivity cannot find the soil layers for you. Apparent resistivity can only give you average soil resistivity numbers from the surface of the earth to whatever spacing the probes can reach. Only a computer-generated soil model has the ability to provide valid soil resistivity numbers.

In practical terms, apparent resistivity uses a single data point (from the four-point test) to provide an average resistivity from the surface of the earth down to a depth roughly equivalent to the probe spacings. Soil modelling compares all of the data points generated from the varied probe spacings (from the four-point test), and compares and evaluates all of them, in order to determine the changes in soil resistivity at depth.

This equation is used to calculate the apparent resistivity of soil/earth for a given probe spacing. You will need to know several bits of information from the Wenner four-point field test: the length of the depth the probes are driven into the earth and the spacings of the probes, and the measured resistance.

$$\rho_a = \frac{4\pi AR}{1 + \dfrac{2A}{\sqrt{(A^2 + 4B^2)}} - \dfrac{2A}{\sqrt{4A^2 + 4B^2}}}$$

where ρ_a = the apparent resistivity of the soil in ohm-metres, R = measured soil resistance in Ω, $\pi = 3.142$, A = spacing of the probes in metres, and B = depth of the probes in metres.

Simplified Apparent Resistivity Equation.   There are simplified versions of the apparent resistivity equation that are useful when the spacing of the probes is greater than 20 times the depth of the testing probes.

The formula for probes spaced in metres is as follows:

Note: One ohm-meter is 3.281 ohm-feet (Ωft).

$$\rho_a = 2\pi AR$$

which can be broken down into:

$$\rho_a = 6.283\ AR$$

where ρ_a = the apparent resistivity of the soil in ohm-metres, $\pi = 3.142$, A = spacing of the probes in metres, and R = measured soil resistance in Ω.

The formula for probes spaced in feet is as follows:

$$\rho_a = 1.915\ AR$$

where ρ_a = the apparent resistivity of the soil in ohm-metres, A = spacing of the probes in feet, and R = measured soil resistance in Ω.

You can easily convert ohm-metres to ohm-centimetres by simply multiplying ohm-meter by 100 to get ohm-centimetres.

Ohm-metres × 100 = ohm-centimetres

Ohm-centimetres / 100 = ohm-metres

Note: Apparent resistivity only looks at a single soil datum to assume a uniform soil for a given spacing. Soil modelling, on the other hand, looks at all of the soil data and compares them in order to find changes in resistivity with depth.

Warning: You do not want to use apparent resistivity for your electrical model, only numbers from soil models should be used.

Simplified Apparent Resistivity Equation with Units. This simplified version of the apparent resistivity equation includes the units, and is useful when dealing with non-typical units. This formula assumes that the spacing of the probes is greater than 20 times the depth of the testing probes.

In the United Kingdom this formula is the most common:

$$\rho_a[\Omega m] = 6.283\,A[m]R[\Omega]$$

where ρ_a = the apparent resistivity of the soil in ohm-metres, A = spacing of the probes in metres, and R = measured soil resistance in Ω.

In the United States this formula is most common:

$$\rho_a[\Omega m] = 1.915\,A[ft]R[\Omega]$$

where ρ_a = the apparent resistivity of the soil in ohm-metres, A = spacing of the probes in feet, and R = measured soil resistance in Ω.

Soil Modelling. Soil modelling is a process of using raw soil resistivity measurements taken during a field test, such as the Wenner four-point method, and processing the data to determine equivalent earth structure models. Computer-generated soil models produce a soil structure with depth of soil layers, the actual soil resistivities at those depths, and soil permittivity values (sometimes given as reflection coefficient and resistivity contrast ratio).

Soil models can be developed using a number of different least-square minimization algorithms: steepest-descent, Levenberg–Marquardt, Fletcher–Powel, G-conjugate, conjugate gradients, and simplex to name a few. The most common algorithm used in developing soil models is typically based on the steepest-descent method and then run through a set of high-precision digital filter coefficients in order to compute the final soil resistivities.

Typical data from a soil model will look something like this (example soil model).

Layer Number	Resistivity (Ωm)	Thickness (m)	Reflection Coefficient (p.u.)	Resistivity Contrast Ratio
1 (air)	Infinite	Infinite	0.0	1.0
2	1037.555	1.343781	−1.0000	0.10376E-16
3	278.4323	15.39296	−0.57685	0.26835
4	1268.895	16.22874	0.64011	4.5573
5	16.52279	Infinite	−0.97429	0.13021E-01

Note how the soil model shows a 1038-Ωm soil going to a depth of 1.3 m, then it changes to 278-Ωm soil for the next 15.4 m (from 1.3 to 16.7 m in depth), and so on. These accurate changes in resistivity at depth are what differentiate the soil model from simple apparent resistivity numbers, which only provides averages from the surface of the earth to depth.

It is very common to see soil models with three (four with air) or more layers. In fact, studies have shown that with the exception of parts of the Gobi Desert in Mongolia, there are no places on the planet that can accurately be modelled as uniform (single layer) soil, and only a few that can be modelled as two layers. For habited locations, the minimum number of layers is three, with an average of four to five layers. Keep in mind that climates with a freezing environment will add layers during the winter as the frost line will increase the soil resistivity by at least a 10× factor.

Hand calculating a valid soil model is simply not practical, given the complexity of the algorithms and digital filter coefficients that are required to generate such a model. However, if you have no other means than hand calculations, you should refer to IEEE Std. 81-2012 Appendix B for more information.

Keep in mind if you are dealing with a grounding electrode system that will ultimately be used to protect personnel in high-voltage environments (human safety), such as eliminating step and touch voltage hazards, your country may have legal requirements (such as 29 CFR 19190.269 in the United States) mandating that services of an engineering firm specializing in designing grounding/earthing systems be used.

Soil Permittivity. Soil Permittivity is a very important engineering factor required for properly calculating the impedance of a grounding system, as the electrode-to-earth interface will be a primary component in how efficiently electric fields will form. Soil permittivity is sometimes provided in the form of a reflection coefficient (p.u.) and a resistivity contrast ratio.

Permittivity: is the measure of the resistance that is encountered when forming an electric field in a medium. In other words, permittivity relates to a material's ability to transmit (or "permit") an electric field.

Measuring the actual permittivity of soil/earth is very new field of geological science and is still considered by many to be more of an academic exercise than a proven test method. As such, you will almost certainly only have the option of using a computer modelling program to perform the calculations needed in order to obtain the actual permittivity number.

However, most soil conditions have calculated permittivities that typically range between 4 and 80, primarily due to soil moisture content. Studies have shown that electrical systems with frequencies lower than 1000 rad/s (159 Hz) are generally not affected by changes in soil permittivity. In other words, if you are designing a grounding system for 60- or 50-Hz systems, any number between 4 and 80 will provide similar results; we recommend using 10 if you do not have the ability to calculate the actual number.

With that being said, if you are designing for a lightning protection system, differing soil permittivity values can impact the impedance of your grounding system by factors greater than 200× at frequencies as low as 1 MHz. In other words, for a 1-MHz signal entering your grounding system, a soil permittivity of 5 can result in an overall impedance to ground of 200× more than if your soil actually has a permittivity of 80.

Soil Layer Interfaces and Its Impact on Grounding Electrode Equations. As we have seen from the section on soil models, standard 3-m (10-ft) grounding electrodes can easily pass through two, three, or more different soil resistivity layers. The interface between these layers is a particular mathematical problem for accurately calculating the resistance/impedance of a grounding system. While we will not cover the equations in this section, as they can only realistically be dealt with using a computer, it is important to understand that these soil interface layers, especially where there are dramatic changes in resistivity from one layer to the next, can generate additional impedances not shown in the following equations.

Resistance-to-Ground and Impedance-to-Ground Testing in the Field. There are several test methods that can be used to conduct actual field measurements of the RTG and/or the impedance to ground of an electrode system or ground grid. These tests, like any other electrical measurement method, can be conducted properly or improperly. Factors such as the training level of the technician conducting the test, the physical arrangements of the grounding system at the site, and the ability to properly isolate electrode systems from other conductive components, all affect the quality of the final test measurements. The most common of these tests is called the three-point fall-of-potential method, and depending on the reference signal and measurement gear used to conduct the test, it can be used to measure both resistance and/or impedance of a grounding/earthing electrode system (ground grid).

The primary goal of most of these calculations in this section is to estimate in advance what the resistance/impedance to ground of the electrode system or grounding grid will be, so that when a field test is conducted, the system meets the required minimal standards.

CALCULATING THE ESTIMATED RESISTANCE TO GROUND OF VERTICAL GROUNDING/EARTHING ELECTRODES

There are many types of vertical grounding (earthing) electrodes: standard-driven ground rods, electrolytic electrodes, vertically installed ground plates, and more. The typical grounding installation will almost always involve at least one standard ground rod.

The calculation of a single ground rod makes for an important example of the complexity involved in calculating the impedance of a single simple electrode in uniform soil conditions. When dealing with a complex grounding electrode system in multilayer soil, calculating the impedance of the system without the aid of a computer is nearly impossible to do especially given the additional calculation issues resulting from multilayer soils and the soil-layer interface.

All of the following calculations are for uniform soil (homogenous/single-layer) conditions.

Single Ground Rod. Calculating the estimated resistance of a single grounding electrode is an important equation for any electrical engineer, especially if you are trying to meet the requirements found in the National Electrical Code 250.53(A)(2) for single ground rod installations.

Presented in this section are two sets of equations. The first equation is for calculating the DC (0 Hz) resistance of a single ground rod in uniform soil. The second equation is for calculating the AC impedance of a single ground rod.

Resistance Equation for a Single Ground Rod. This single equation is used to calculate the DC resistance of a single ground rod in uniform soil. You need to know the length and radius of the ground rod, and the soil resistivity of where it will be installed.

$$R = \frac{\rho}{2\pi L}\left(\ln\frac{4L}{r} - 1\right)$$

Typically electrodes are provided with a diameter, this formula uses radius. If you are given a diameter of the electrode, you must divide that number by two in order to get a radius.

where R = the calculated resistance of the electrode in Ω, ρ = the resistivity of the soil in ohm-centimetres, π = 3.142, r = radius of the electrode in centimetres, L = length of electrode in centimetres, and ln = natural log function.

If we take a 10 ft × ¾ in. ground rod in 100-Ωm soil, we first need to convert all the units to centimetres. So our ground rod would be 304.8 cm × 1.905 cm in 10000-Ωcm soil. Our ground rod has a diameter of 1.905 cm, which divided by 2 would be a radius of 0.9525 cm. Thus we get the following equation:

$$R = \frac{10,000}{2\pi 304.8}\left(\ln\frac{4\times304.8}{0.9525} - 1\right)$$

If we solve the multiplication we get:

$$R = \frac{10,000}{1915.11}\left(\ln\frac{1219.2}{0.9525} - 1\right)$$

If we solve the division we get:

$$R = 5.22(\ln 1280 - 1)$$

If we solve the natural log we get:

$$R = 5.22(7.1546 - 1)$$

We can then solve the complete equation which gives us this final result in ohms:

$$R = 32.13$$

In BS 7430:2011+A1:2015 Chapter 9.5.3, this equation given for a single rod electrode uses diameter instead of radius, which results in the following equation:

$$R_r = \frac{\rho}{2\pi L}\left(\log_e\frac{8L}{d} - 1\right)$$

where R_r = the calculated resistance of the electrode (rod) in Ω, ρ = the resistivity of the soil in Ωm (ohm-metres), π = 3.142, d = diameter of the electrode in metres, L = length of electrode in metres, and $\log_e$ = natural log function.

Note: A change in the diameter of the rod has little effect on the overall resistance-to-earth of the electrode. Increases in diameter typically are done to improve the mechanical strength of the rod electrode when being forcefully driven into the earth.

Impedance Equation for a Single Ground Rod.   This six-step equation is used to calculate the AC impedance of a single ground rod in uniform soil. When we have a single ground rod installed for an AC system, 50 or 60 Hz, we must take into account at least three more factors: leakage conductance (G), capacitance (C), and inductance (L).

Note: Unless you are modelling lightning or other high-frequency noise, 50-/60-Hz systems have a frequency that is low enough that you can generally model them using the DC equations.

It should be noted that the formulas presented here are simplified impedance equations, as they do not take into consideration the material properties of the electrode itself. As an example, the difference in impedance between a copper electrode and a low-carbon steel electrode can be quite significant for high-frequency events. Computer modelling software will be required to adequately calculate the differences in impedance between material properties of differing electrodes.

The impedance equation for a single ground rod presented here is based on standard telegraphers' equations (or transmission line equations) and is represented in Fig. 8.3 (image courtesy of Wikipedia). The resistance of the electrode itself is very small compared to the other terms in the equation, it is therefore ignored.

You need to know the length of the ground rod, the diameter of the ground rod, the soil resistivity of where it will be installed, the permittivity of the soil, the frequency of your electrical system in radians per second, and know several mathematical constants (provided below).

Note: It is important that all units in the following six-step formulas are the same. In this case, all units are presented in metres. It is important that you convert any feet or centimetre units to metres for these formulas.

Step 1: Leakage Conductance (G)   This equation is used to calculate the ease in which current will be able to leave the electrode and "leak" into the surrounding soil/earth. Now, you will need to know the length of the electrode, the diameter of the electrode, and the resistivity of the uniform soil. The result of this formula will be used later on in Steps 4 and 5.

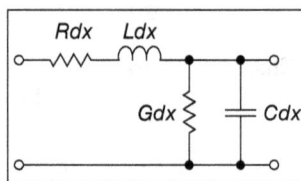

Figure 8.3 Telegraphers' equation.

Note: This formula requires soil resistivity to be provided in ohm-metres and not ohm-centimetres.

$$G = \frac{2\pi}{\rho \ln \frac{4l}{d}}$$

where G = siemens (Mho) per meter (S/m), l = length of the electrode in metres, d = diameter of the electrode in metres, ρ = the resistivity of the soil in ohm-metres, π = 3.142, and ln = natural log function.

Step 2: Capacitance (C)   This equation is used to calculate the ability of the electrode to store an electrical charge. You will need to know the length of the electrode, the diameter of the electrode, the resistivity of the uniform soil, and the permittivity of the uniform soil. The result of this formula will be used later on in Steps 4 and 5.

If you have the relative permittivity of your soil from your computer-generated soil model report, usually a number between 4 and 80, please use if for ε_r. However, if you do not have the ability to calculate soil permittivity, and your electrical system is being used for a 50-/60-Hz electrical system, we recommend setting $\varepsilon_r = 10$, as a lower number is more conservative. Please see the section on soil resistivity for more information regarding soil permittivity.

It is recommended that you use metres for this formula, as the constant for the permittivity of free space is in farads per meter.

$$C = \frac{2\pi \varepsilon_r \varepsilon_o}{\ln \frac{4l}{d}}$$

where C = farads per meter (F/m), ε_r = relative permittivity of the soil (epsilon) (typically between 4 and 80), ε_0 = permittivity of free space (constant = 8.854×10^{-12} F/m), l = length of the electrode in metres, d = diameter of the electrode in metres, π = 3.142, and ln = natural log function.

Step 3: Inductance (L)   This equation is used to calculate the amount of electromotive force (voltage) that the change in current flowing in the electrode will induce into both itself (self-inductance) and the adjacent soil (mutual inductance). You will need to know the length of the electrode, and the diameter of the electrode. The result of this formula will be used later on in Steps 4 and 5.

$$L = \frac{\mu_0}{2\pi} \ln \frac{4l}{d}$$

where L = henries per meter (H/m), μ_0 = permeability constant $4\pi \times 10^{-7}$ (0.000001257) in H/m, l = length of the electrode in metres, d = diameter of the electrode in metres, π = 3.142, and ln = natural log function.

Note: This equation assumes $\mu_r = 1$, so it has been removed from the equation. However, if you know μ_r you can add it back into the formula by simply multiplying it by μ_0.

Step 4: Wave Propagation Parameter (α)   This equation is based on standard telegraphers' equations, and is used to calculate how effectively electromagnetic waves will propagate through the electrode/soil interface. You will need to know the values from Steps 1, 2, and 3, and the frequency in radians per second of your electrical system. The result of this formula will be used later on in Steps 5 and 6.

$$\alpha = \sqrt{jwL(G + jwC)}$$

where α = wave propagation parameter, j = imaginary number (square root of -1), w = frequency in radians per second (60 Hz = 377/50 Hz = 314), G = leakage conductance in S/m from Step 1, C = capacitance in F/m from Step 2, and L = inductance in H/m from Step 3.

Step 5: Characteristic Impedance (Z_c)   This equation is used to calculate the effective impedance of the electrode. You will need to know the values from Steps 1, 2, and 3, and the frequency in radians per second of your electrical system. The result of this formula will be used later on in Step 6.

$$Z_c = \sqrt{\frac{jwL}{G + jwC}}$$

where Z_c = characteristic impedance, j = imaginary number (square root of -1) w = frequency in radians per second (60 Hz = 377/50 Hz = 314), G = leakage conductance in S/m from Step 1, C = capacitance in F/m from Step 2, and L = inductance in H/m from Step 3.

Step 6: Impedance Equation for a Single Ground Rod   This final equation is used to calculate the effective impedance to ground of the electrode. You will need to know the values from Steps 1, 2, and 3, and the frequency in radians per second of your electrical system.

$$Z = \frac{e^{2\alpha l} + 1}{e^{2\alpha l} - 1} Z_c$$

where Z = the impedance of the single ground rod, Z_c = characteristic impedance from Step 5, e = Euler's number (constant = 2.718), α = wave propagation parameter from Step 4, and l = length of the electrode in metres.

Current Capacity (Ampacity) of a Single Ground Rod.   The ability of a grounding electrode to handle current (ampacity) is largely dependent on the resistivity soil surrounding the electrode, and particularly the moisture content of the surrounding soil.

For homogeneous (uniform) soil conditions, it has been estimated that 25% of a ground rods resistance occurs within the top 0.03 m radius of the surface of the rod, approximately 3 cm. This means that the top section of a ground rod can get hot enough to actually boil (vaporize) the water out of the soil, resulting in what is called a "smoking" ground rod.

However, this does not hold true for common multilayer soil conditions, where ground rods can have a very different resistance profiles. For high-over-low soil conditions, approximately 25% of the ground rods resistance will occur within the top 15 cm of rod. For low-over-high soil conditions, there is simply no generalization that can be made at all about the resistance profile of a ground rod.

The following equation will provide a not to exceed (maximum) current per electrode foot for a single ground rod.

$$I = \frac{34800 \times d \times L}{\sqrt{\rho \times t}}$$

where I = the calculated amps per meter of electrode, ρ = the resistivity of the soil in ohm-metres, L = length of the rod (below grade) in metres, d = diameter of the electrode in metres, and t = current duration in seconds.

Simplified Calculation for Single Ground Rod. The following formulas are standard equations from the IEEE Std. 142 Green Book and are approximations that come within 2% of the full ground rod equation formula section B.a.i.1.

10 ft (3 m) by ½ in Diameter Standard Ground Rod. This simplified equation is used to calculate the DC RTG of a single 10 ft × ½ in. ground rod in uniform soil. You only need to know the uniform (homogenous) soil resistivity for this calculation.

$$R = \frac{\rho}{288}$$

where R = the calculated resistance of the electrode in Ω and ρ = the resistivity of the soil in ohm-centimetres.

10 ft (3 m) by 5/8 in Diameter Standard Ground Rod. This simplified equation is used to calculate the DC RTG of a single 10 ft × 5/8 in. ground rod in uniform soil. You only need to know the uniform soil resistivity for this calculation.

$$R = \frac{\rho}{298}$$

where R = the calculated resistance of the electrode in Ω and ρ = the resistivity of the soil in ohm-centimetres.

10 ft (3 m) by ¾ in Diameter Standard Ground Rod. This simplified equation is used to calculate the DC RTG of a single 10 ft × ¾ in. ground rod in uniform soil. You only need to know the uniform soil resistivity for this calculation.

$$R = \frac{\rho}{307}$$

where R = the calculated resistance of the electrode in Ω and ρ = the resistivity of the soil in ohm-centimetres.

Two Ground Rods with Spacings Less Than the Length of the Rods. This equation is used to calculate the DC RTG of two similar ground rods of any length that are connected together above ground, spaced apart at a distance that is less than the length of the ground rods, and in uniform soil. You need to know the uniform soil resistivity, the length of the electrodes (they must be the same length), and the distance they are apart from each other, and the radius of the ground for this calculation.

At its root, this formula assumes that each ground rod will utilize a cylinder of soil (sphere of influence) and that the flow of electricity to ground will act as a dielectric flux from an isolated charged cylinder. This means that the RTG of the two ground rods is essentially the same as the capacitance of isolated cylinders with lengths far greater than their radii. Ultimately, the formula approximates the volume of the soil between the rods in order to get a final estimated resistance of the two electrodes.

Note: This is a simplified formula and there is no reason to continue the formula past the shown mathematical series.

$$R = \frac{\rho}{4\pi L}\left(\ln\frac{4L}{r} - 1\right) + \frac{\rho}{4\pi s}\left(1 - \frac{L^2}{3s^2} + \frac{2L^4}{5s^4}\cdots\right)$$

where R = the calculated resistance of the electrode in Ω, ρ = the resistivity of the soil in ohm-centimetres, π = 3.142, r = radius of the ground rod in centimetres, L = length of the rod (below-grade) in centimetres, s = spacing between the rods in centimetres, and ln = natural log function.

Two Ground Rods with Spacings Greater Than the Length of the Rods. This equation is used to calculate the DC RTG of two similar ground rods of any length that are connected together above ground, spaced apart at a distance that is greater than the length of the ground rods, and in uniform soil. You need to know the uniform soil resistivity, the length of the electrodes (they must be the same length), and the distance they are apart from each other, and the radius of the ground for this calculation.

At its root, this formula assumes that each ground rod will utilize a cylinder of soil (sphere of influence) and that the flow of electricity to ground will act as a dielectric flux from an isolated charged cylinder. This means that the RTG of the two ground rods is essentially the same as the capacitance of isolated cylinders with lengths far greater than their radii. Ultimately, the formula approximates the volume of the soil between the rods in order to get a final estimated resistance of the two electrodes.

Note: This is a simplified formula and there is no reason to continue the formula past the shown mathematical series.

$$R = \frac{\rho}{4\pi L}\left(\ln\frac{4L}{r} + \ln\frac{4L}{s} - 2 + \frac{s}{2L} - \frac{s^2}{16L^2} + \frac{s^4}{512L^4}\cdots\right)$$

where R = the calculated resistance of the electrode in Ω, ρ = the resistivity of the soil in ohm-centimetres, π = 3.142, r = radius of the rod in centimetres, L = length of the rod (below grade) in centimetres, s = spacing between the rods in centimetres, and ln = natural log function.

If you are using ¾-in. diameter ground rods, you can use this simpler equation:

$$R = \frac{\rho}{4\pi L}\left(\frac{6.155}{L} + \frac{1}{S}\right)$$

where R = the calculated resistance of the electrode in Ω, ρ = the resistivity of the soil in ohm-centimetres, π = 3.142, L = Length of the rod (below grade) in centimetres, and s = spacing between the rods in centimetres.

Parallel Connection of Aligned Rods.   In BS 7430:2011+A1:2015 Chapter 9.5.4, this equation given for a multiple rod electrode installed in a linear (aligned) row. This equation uses diameter instead of radius, which results in the following equation:

$$R_t = \frac{1}{n} \times \frac{\rho}{2\pi L}\left[\log_e\left(\frac{8L}{d}\right) - 1 + \frac{\lambda L}{s}\right]$$

where R_t = the calculated resistance of the total of electrodes (rods) in Ω, ρ = the resistivity of the soil in ohm-metres, π = 3.142, d = diameter of the electrode in metres, L = length of electrode in metres, n = number of rods, $\log_e$ = natural log function, and λ = a group factor as shown below:

$$\lambda = 2\Sigma\left(\frac{1}{2} + \cdots + \frac{1}{n}\right)$$

For larger values of n, λ can be approximated by:

$$\lambda \cong 2\log_e \frac{1.781n}{2.718}$$

Multiplying Factors for Multiple Rods.   This DC formula is for use with uniform soil and comes from the IEEE Std. 142 Green Book. When dealing with multiple ground rods aligned at least one rod length apart from each other, arranged in a straight line, hallow triangle, hallow square, or hallow circle, use the formula below to estimate the effect of multiple rods. You will need to calculate the DC resistance of a single ground rod as found above.

$$R_t = \frac{R_1}{E_n} \times F$$

where R_t = resistance total of all rods, R_1 = calculated resistance of a single ground rod (see section above), E_n = number of electrodes, and F = factor from the table below.

Multiplying Factors for Multiple Rods

Number of Rods	Factor
2	1.16
3	1.29
4	1.36
8	1.68
12	1.80
16	1.92
20	2.00
24	2.16

So let us put this formula in action by starting with a single 10 ft × ¾ in. ground rod in uniform 10,000-Ωcm soil (100 Ωm). If we use the simplified formula from the section above, we find that we only need to divide 10,000 Ωcm by 307 to get an estimated resistance of a single ground rod of 32.6 Ω estimated RTG.

If we choose to install a series of four ground rods arranged in a straight line we can plug the numbers into the formula and get:

$$R_t = \frac{32.6}{4} \times 1.36$$

By doing the math we get the total resistance of four ground rods of 11.1 Ω.

If we compare this formula against a computer calculation using 10 ft × ¾ in. ground rods configured in a straight line, spaced 20 ft apart, with the interconnecting conductor bonding the ground rods together above grade, and with a bare conductor buried 1.5-ft below grade (acting as part of the electrode system), we get the following chart (in uniform 100-Ωm soil):

# of Ground Rods	Linear Feet of Copper 4/0 AWG Conductor (ft)	Hand Resistance Estimate using Multiplying Factor (Ω)	Computer Impedance Calculation with Conductor Above Grade (Ω)	Computer Impedance Calculation with Conductor Below Grade (Ω)*
1	0	32.6	29.3	29.3
2	20	18.9	16.3	12.7
3	40	14.0	11.6	8.8
4	60	11.1	9.2	6.9
8	140	6.9	5.1	3.9
12	220	4.9	3.7	2.8
16	300	3.9	2.9	2.2
20	380	3.3	2.4	1.9
24	460	2.9	2.0	1.6

*An actual field measurement will return a result closest to this estimate.

As can be seen in the chart, the hand resistance estimate using the multiplying factor gives us estimates that are much higher than what we would expect to see in the field. For example, if you are trying to design a grounding electrode system that has a resistance/impedance to ground of less than 5 Ω in 100-Ωm uniform soil, the multiplying factor formula will tell you that you need between 12 and 16 ground rods; when in fact you really only need six to seven ground rods (with a buried conductor).

Round Plate Buried Vertically.   This equation is used to calculate the DC RTG of a round plate buried vertically in uniform soil. You need to know the uniform soil resistivity, the radius of the plate electrode, and the depth the plate is buried for this calculation.

Note 1: The depth of the plate is measured from the centre point of the plate.

Note 2: This is a simplified formula and there is no reason to continue the formula past the shown mathematical series.

$$R \doteq \frac{\rho}{8r} + \frac{\rho}{4\pi s}\left(1 - \frac{7r^2}{24s^2} + \frac{99r^4}{320s^4}\cdots\right)$$

where R = the calculated resistance of the electrode in Ω, ρ = the resistivity of the soil in ohm-centimetres, π = 3.142, r = radius of plate in centimetres, and s = depth to the centre of the plate (below grade) in centimetres, multiplied by 2.

Buried Horizontal Electrodes.   There are many different configurations for buried horizontal grounding (earthing) electrodes: ground rings, straight wires, straps, angled wires, several star shapes, and more.

The typical grounding installation will almost always involve at some configuration of a horizontal electrode system, along with at least one standard ground rod. The calculation of a buried horizontal component of your grounding system (along with the calculation of the vertical component) is critical in understanding the overall RTG of your grounding electrode system.

When dealing with a complex grounding electrode system in multilayer soil, calculating the impedance of the system without the aid of a computer is nearly impossible to do, especially given the additional calculation issues resulting from soil-layer interface. When dealing with human safety issues, such as the calculation of step and touch voltages, it would be unethical in my opinion not to use advanced computer programs designed for such analysis.

All of the following calculations are for uniform soil conditions with DC electrical systems. None of these calculations take into account the material properties of the electrode. This means the difference between a copper electrode and an electrode of higher impedance such as steel is not taken into account.

Note 1: Most of the following equations require you to determine the depth the electrode will be installed below grade and multiply that number by 2. This is done as an approximation of the volume of soil that is above and below the electrode, a conservative sphere of influence.

Note 2: The units for the following equations must be in centimetres, because the formulas use constants (approximations) that were calculated assuming centimetres as a base unit.

Straight Wire/Strip of round conductor electrodes.   This equation is used to calculate the DC RTG of a straight wire buried horizontally in uniform soil. You need to know the uniform soil resistivity, the length of the wire electrode, the radius of the wire, and the depth of the wire for this calculation.

Note: This is a simplified formula and there is no reason to continue the formula past the shown mathematical series.

$$R = \frac{\rho}{4\pi L}\left(\ln\frac{4L}{r} + \ln\frac{4L}{s} - 2 + \frac{s}{2L} - \frac{s^2}{16L^2} + \frac{s^4}{512L^4}\cdots\right)$$

where R = the calculated resistance of the electrode in Ω, ρ = the resistivity of the soil in ohm-centimetres, π = 3.142, r = radius of the rod in centimetres, L = Length of the wire in centimetres, s = depth of the wire (below grade) in centimetres, multiplied by 2, and In = natural log function.

In BS 7430:2011+A1:2015 Chapter 9.5.5, this equation given for a strip or round conductor electrode installed in a straight line. This equation uses diameter instead of radius, which results in the following equation:

$$R_{ta} = \frac{\rho}{2\pi L} \log_e \left(\frac{L^2}{\kappa h d} \right)$$

where R_{ta} = the calculated resistance of the strip or round conductor electrodes in Ω, ρ = the resistivity of the soil in ohm-metres, π = 3.142, d = the width of the strip or the diameter of the round conductor in metres, L = length of the strip or conductor in metres, h = depth of the electrode in metres, $\log_e$ = natural log function, and κ = has the value of 1.36 for a strip and 1.83 for a round conductor.

When two or more straight conductors (as per above) are installed parallel to each other and separated by a fixed parallel distance (in metres), the combined resistance can be calculated as follows:

$$R_n = FR_1$$

where R_n = the calculated resistance of n conductors in parallel in Ω, R_1 = the resistance of a single length of electrode from the equation above (R_{ta}) in ohms (Ω), and F = the following values provided that $0.02 < (s/L) < 0.3$:

For two lengths: $F = 0.5 + 0.078(s/L)^{-0.307}$
For three lengths: $F = 0.33 + 0.071(s/L)^{-0.408}$
For four lengths: $F = 0.25 + 0.067(s/L)^{-0.451}$

where s = the spacing between electrodes in metres, and L = the length of the electrodes in metres.

Right-Angle Wire.   This equation is used to calculate the DC RTG of a wire with a right angle of equal lengths buried horizontally in uniform soil. You need to know the uniform soil resistivity, the length of the wire electrode, the radius of the wire, and the depth of the wire for this calculation.

Note: This is a simplified formula and there is no reason to continue the formula past the shown mathematical series.

$$R = \frac{\rho}{4\pi L} \left(\ln \frac{2L}{r} + \ln \frac{2L}{s} - 0.2373 + 0.2146 \frac{s}{L} + 0.1035 \frac{s^2}{L^2} - 0.0424 \frac{s^4}{L^4} \cdots \right)$$

where R = the calculated resistance of the electrode in Ω, ρ = the resistivity of the soil in ohm-centimetres, π = 3.142, r = radius of the rod in centimetres, L = length of the wire in centimetres, s = depth of the wire (below grade) in centimetres, multiplied by 2, and In = natural log function.

Three-Point Star or "Y".   This equation is used to calculate the DC RTG of a wire-type electrode with three arms of equal lengths, buried horizontally in uniform soil and configured into the shape of a "Y" or three-pointed star. You need to know the uniform soil resistivity, the length of the one arm of the wire electrode, the radius of the wire, and the depth of the wire for this calculation.

Note: This is a simplified formula and there is no reason to continue the formula past the shown mathematical series.

$$R = \frac{\rho}{6\pi L}\left(\ln\frac{2L}{r} + \ln\frac{2L}{s} + 1.071 - 0.209\frac{s}{L} + 0.238\frac{s^2}{L^2} - 0.054\frac{s^4}{L^4}\cdots \right)$$

where R = the calculated resistance of the electrode in Ω, ρ = the resistivity of the soil in ohm-centimetres, π = 3.142, r = radius of the rod in centimetres, L = length of the one arm of the wire electrode in centimetres, s = depth of the wire (below grade) in centimetres, multiplied by 2, and $\ln$ = natural log function.

VERTICAL RODS IN A HOLLOW SQUARE (EARTHING RING)

In BS 7430:2011+A1:2015 Section 9.5.8.5, page 43, we find the formula for calculating the resistance to earth of a buried square ring (hollow square) with earth rods. This formula allows you to pick the number of rods per a given side and then calculate the total resistance (R_{TOT}). The spacings of the rods and overall number of rods automatically determines the size of the square. Here is the formula:

$$R_{TOT} = R_r\left(\frac{1+\lambda a}{N}\right)$$

where

$$a = \frac{\rho}{2\pi R_r S}$$

where the shape is a square:

$$N = 4(n-1)$$

And where R_r = the resistance of a single electrode in ohms (Ω), λ = the factor in Table 2 below, ρ = the resistivity of the soil in ohm metres, π = 3.142, S = spacing of the rods in metres, N = the number of rods used as electrodes.

If we were to do an example of a square ground ring with eight earth rods (3 m long) spaced at 6-m intervals (see Fig. 8.4), with each individual earth rod measuring at 35 ohms, and a soil resistivity of 100 Ωm, we can start to figure out our formula.

Therefore, we start by figuring out for N, which gives us the following:

$$N = 4(n-1)$$

We know that we have 3 rods along the side of the square, so therefore $n = 3$, which gives us the following:

$$N = 4(3-1)$$

Table 2 (page 43)

BS 7430:2011+A1:2015	
Table 2 (Page 43)	
Factors for vertical elecrodes arranged in a hollow square (earthing ring)	
Number of Electrodes (n) Along the Side of the Square	Factor λ
2	2.71
3	4.51
4	5.46
5	6.14
6	6.63
7	7.03
8	7.30
9	7.65
10	7.90
12	8.22
14	8.67
16	8.95
18	9.22
20	9.40

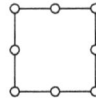

Figure 8.4 Square earthing ring with 8 earth rods.

When we solve for N we find that $N = 8$. This also happens to be the total number of rods in our square.

Now, we need to solve for "a" using the formula from above:

$$a = \frac{\rho}{2\pi R_r S}$$

We know that our soil resistivity (ρ) = 100, that $2\pi = 6.283$, and that our individual earth rods (R_r) measure = 35, and or spacing of the rods (S) = 6. So therefore, we get the following:

$$a = \frac{100}{(6.283)(35)(6)}$$

Therefore, $a = 0.076$

We can now build the following formula:

$$R_{TOT} = R_r \left(\frac{1 + \lambda a}{N} \right)$$

Since we know that we have three rods on the side of our square we can look on Table 2 to find our factor $\lambda = 4.51$. We know that $N = 8$ from our equation above, that $a = 0.076$ from our equation above, and that our resistance of a single earth rod (R_r) = 35. Therefore, we can build the following:

$$R_{TOT} = 35 \left(\frac{1 + 4.51(0.076)}{8} \right)$$

which gives us $R_{TOT} = 5.8745$ ohms.

If we calculate this same system using the CDEGS software program, we get a calculated resistance of 3.98 ohms. Meaning that this formula tends to be overly conservative in its resulting answers by a factor of at least 1.5×.

VERTICAL RODS IN A HOLLOW RECTANGLE (EARTHING RING)

In BS 7430:2011+A1:2015 Section 9.5.8.5, page 43, we find the formula for calculating the resistance to earth of a buried rectangular ring (hollow rectangle) with earth rods. This formula allows you to pick the number of rods per a given side and then calculate the total resistance (R_{TOT}). The spacings of the rods and overall number of rods automatically determines the size of the square. Here is the formula:

$$R_{TOT} = R_r \left(\frac{1 + \lambda a}{N} \right)$$

where

$$a = \frac{\rho}{2 \pi R_r S}$$

where the shape is a rectangle (where N is the total number of electrodes):

$$n = \frac{N}{4} + 1$$

And where R_r = the resistance of a single electrode in ohms (Ω), λ = the factor in Table 2 below, ρ = the resistivity of the soil in ohm metres, $\pi = 3.142$, S = spacing of the rods in metres, N = the number of rods used as electrodes.

Note from BS 7430: This formula will provide an error rate of less than 6%, if the length to width ratio of the rectangle does not exceed a factor of two.

If we were to do an example of a rectangular ground ring with 10 earth rods (3 m long) spaced at 6-m intervals (see Fig. 8.5), with each individual earth rod measuring at 35 ohms, and a soil resistivity of 100 Ωm, we can start to figure out our formula.

BS 7430:2011+A1:2015	
Table 2 (Page 43)	
Factors for vertical elecrodes arranged in a hollow square (earthing ring)	
Number of Electrodes (n) Along the Side of the Square	Factor λ
2	2.71
3	4.51
4	5.46
5	6.14
6	6.63
7	7.03
8	7.30
9	7.65
10	7.90
12	8.22
14	8.67
16	8.95
18	9.22
20	9.40

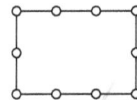

Figure 8.5 Rectangular earth ring with 10 earthing rods.

Because we have 10 earth rods, we know that N is the total number of electrodes that make $N = 10$. However, we have a problem in figuring out what n should be, because we have four rods on one side of the rectangle, and three rods on the other side... so which do we choose? This is why we have the following formula that will tell us what to n should be.

$$n = \frac{N}{4} + 1$$

Note: This formula only works for equally spaced earth rods.
If we insert $N = 10$, we get the following:

$$n = \frac{10}{4} + 1$$

When we solve for n we find that $n = 3.5$ which rounds up to 4. We can now look at Table 2 above and find that $\lambda = 5.46$.

Now, we need to solve for "a" using the formula from above:

$$a = \frac{\rho}{2\pi R_r S}$$

We know that our soil resistivity $(\rho) = 100$, that $2\pi = 6.283$, and that our individual earth rods (R_r) measure $= 35$, and or spacing of the rods $(S) = 6$. So therefore, we get the following:

$$a = \frac{100}{(6.283)(35)(6)}$$

Therefore, $a = 0.076$

We can now build the following formula:

$$R_{TOT} = R_r \left(\frac{1 + \lambda a}{N} \right)$$

Since we know that $n = 4$ from the calculation above, we can look on Table 2 to find our factor $\lambda = 5.46$. We know that $N = 10$ because this is the total number of rods in our rectangle, we also know that $a = 0.076$ from our equation above, and that our resistance of a single earth rod $(R_r) = 35$. Therefore, we can build the following:

$$R_{TOT} = 35 \left(\frac{1 + (5.46)(0.076)}{10} \right)$$

Which gives us $R_{TOT} = 4.95$ ohms

If we calculate this same system using the CDEGS software program, we get a calculated resistance of 3.33 ohms. Meaning that this formula tends to be overly conservative in its resulting answers by a factor of around 1.5×.

Four-Point Star or "+". This equation is used to calculate the DC RTG of a wire-type electrode with four arms of equal lengths buried horizontally in uniform soil and configured into the shape of a "+," cross, or four-pointed star. You need to know the uniform soil resistivity, the length of the one arm of the wire electrode, the radius of the wire, and the depth of the wire for this calculation.

Note: This is a simplified formula and there is no reason to continue the formula past the shown mathematical series.

$$R = \frac{\rho}{8\pi L} \left(\ln \frac{2L}{r} + \ln \frac{2L}{s} + 2.912 - 1.071 \frac{s}{L} + 0.645 \frac{s^2}{L^2} - 0.145 \frac{s^4}{L^4} \cdots \right)$$

where R = the calculated resistance of the electrode in Ω, ρ = the resistivity of the soil in ohm-centimetres, $\pi = 3.142$, r = radius of the rod in centimetres, L = length of the one arm of the wire electrode in centimetres, s = depth of the wire (below grade) in centimetres, multiplied by 2, and In = natural log function.

Six-Point Star. This equation is used to calculate the DC RTG of a wire-type electrode with six arms of equal lengths buried horizontally in uniform soil and configured into the shape of a six-pointed star. You need to know the uniform soil resistivity, the length of one arm of the wire electrode, the radius of the wire, and the depth of the wire for this calculation.

Note: This is a simplified formula and there is no reason to continue the formula past the shown mathematical series.

$$R = \frac{\rho}{12\pi L}\left(\ln\frac{2L}{r} + \ln\frac{2L}{s} + 6.851 - 3.128\frac{s}{L} + 1.758\frac{s^2}{L^2} - 0.490\frac{s^4}{L^4}\cdots\right)$$

where R = the calculated resistance of the electrode in Ω, ρ = the resistivity of the soil in ohm-centimetres, π = 3.142, r = radius of the rod in centimetres, L = length of the one arm of the wire electrode in centimetres, s = depth of the wire (below grade) in centimetres, multiplied by 2, and $\ln$ = natural log function.

Eight-Point Star.   This equation is used to calculate the DC RTG of a wire-type electrode with eight arms of equal lengths buried horizontally in uniform soil and configured into the shape of an eight-pointed star or eight-pointed asterisk. You need to know the uniform soil resistivity, the length of the one arm of the wire electrode, the radius of the wire, and the depth of the wire for this calculation.

Note: This is a simplified formula and there is no reason to continue the formula past the shown mathematical series.

$$R = \frac{\rho}{16\pi L}\left(\ln\frac{2L}{r} + \ln\frac{2L}{s} + 10.98 - 5.51\frac{s}{L} + 3.26\frac{s^2}{L^2} - 1.17\frac{s^4}{L^4}\cdots\right)$$

where R = the calculated resistance of the electrode in Ω, ρ = the resistivity of the soil in ohm-centimetres, π = 3.142, r = radius of the rod in centimetres, L = length of the one arm of the wire electrode in centimetres, s = depth of the wire (below grade) in centimetres, multiplied by 2, and $\ln$ = natural log function.

Straight Strap.   This equation is used to calculate the DC RTG of a straight strap buried horizontally in uniform soil. You need to know the uniform soil resistivity, the length of the wire electrode, the thickness of the strap, the width of the strap, and the depth of the wire for this calculation.

Note: This is a simplified formula and there is no reason to continue the formula past the shown mathematical series.

$$R = \frac{\rho}{4\pi L}\left(\ln\frac{4L}{a} + \frac{a^2 - \pi ab}{2(a+b)^2} + \ln\frac{4L}{s} - 1 + \frac{s}{2L} - \frac{s^2}{16L^2} + \frac{s^4}{512L^4}\cdots\right)$$

where R = the calculated resistance of the electrode in Ω, ρ = the resistivity of the soil in ohm-centimetres, π = 3.142, a = thickness of strap in centimetres, b = width of strap in centimetres, L = length of strap in centimetres, and s = depth of the strap (below grade) in centimetres, multiplied by 2, $\ln$ = natural log function.

Circular Ring.   This equation is used to calculate the DC RTG of a circular ring of wire buried horizontally in uniform soil. You need to know the uniform soil resistivity, the diameter of the wire electrode, the diameter of the wire, and the depth of the wire for this calculation.

$$R = \frac{\rho}{2\pi^2 D}\left(\ln\frac{8D}{d} + \ln\frac{4D}{s}\right)$$

where R = the calculated resistance of the electrode in Ω, ρ = the resistivity of the soil in ohm-centimetres, π = 3.142, D = diameter of the ground ring in centimetres, d = diameter of the wire in centimetres, s = depth of the ring (below grade) in centimetres, multiplied by 2, and In = natural log function.

Round Plate Buried Horizontally. This equation is used to calculate the DC RTG of a round plate buried horizontally in uniform soil. You need to know the uniform soil resistivity, the radius of the plate electrode, and the depth the plate is buried for this calculation.

Note: This is a simplified formula and there is no reason to continue the formula past the shown mathematical series.

$$R = \frac{\rho}{8r} + \frac{\rho}{4\pi s}\left(1 - \frac{7r^2}{12s^2} + \frac{33r^4}{40s^4}\cdots\right)$$

where R = the calculated resistance of the electrode in Ω, ρ = the resistivity of the soil in ohm-centimetres, π = 3.142, r = radius of plate in centimetres, and s = depth of the plate (below grade) in centimetres, multiplied by 2.

In BS 7430:2011+A1:2015 Chapter 9.5.2, which results in the following equation:

$$R = \frac{\rho}{4}\sqrt{\frac{\pi}{A}}$$

where R = the calculated resistance of the plate electrode in Ω, ρ = the resistivity of the soil in ohm metres, π = 3.142, A = the area of one face of the plate, in square metres (m²)

Note: For calculating multiple plate electrodes, the resistance is roughly inversely proportional to the linear dimensions of the plates strung together, and not to the combined surface areas of the plates.

Calculating Electrodes in Multilayer Soil Models. The above equations have shown how to calculate a variety of grounding electrode configurations in theoretical uniform (or homogenous) soil conditions. In practice, it is rare to find soil conditions of less than three layers. Other factors such as rain, dry summers, and frozen ground in winter can actually add additional soil layers from one season to another. It is important to understand the differences that multilayer soils can have on the overall resistance/impedance of an electrode.

One of the best methods for calculating the effects of multilayer soils is the finite element method (FEM), which is best applied to a two-layer soil model, but can be applied to multilayer soils. The FEM takes the approach that the sphere of influence defines a volume of soil utilized by the grounding electrode, and that by breaking down the soil surrounding the electrode into dozens of individual triangles, you can then analyze each individual triangle, in its corresponding soil, to obtain a final resistance/impedance for the electrode. The method is tedious as the more soil layers one must analyze, the number of triangles increases exponentially. The FEM approach also does not take into account the impact that the soil layer interface will have on the electrode, but the method is one of the most accurate hand-calculation methodologies we currently have at our disposal. Some white papers report accuracy rates of less than 5% using the FEM approach.

Unfortunately, it would take an entire section just to explain this methodology, which we simply do not have the time or space in this book to do. Please see the Bibliography section for more information on published white papers on the FEM approach.

ESTIMATING THE RESISTANCE/IMPEDANCE OF COMPLEX GROUNDING ELECTRODE SYSTEMS

It is very common for electrical engineers to need a way to estimate the overall resistance/impedance to ground of a grounding grid. It should be clear by now, calculating each individual ground rod and every conductor in a grid would be a very tedious task. Fortunately, there are a few simple formulas that can be used for getting a rough estimate of a ground grid.

Simplified Resistance Calculations for Ground Grids or Mesh. To get a rough estimate of what the DC RTG of your complex grounding/earthing grid will be, even in multilayer soils, the following simple formula is a great way to get it. This formula will work for nearly any size ground grid of any shape, as long as the entire grounding/earthing grid has similar soil conditions.

$$R = \frac{\rho}{4r}$$

where R = the calculated resistance of the electrode in Ω, ρ = the resistivity of the soil in ohm-metres, and r = equivalent grid radius in metres.

Step 1: You will only need to know the resistivity of the bottom (or infinite) layer soil from your soil model, as the bottom layer will be the primary driver for determining the RTG of large ground grids. In our soil model example in the Soil Model section, we can see that our bottom layer soil is 16.52 Ωm. We will call this 17-Ωm soil for simplicity sakes.

Step 2: This formula is based on the area of a circle. So for the next step, you will need to convert the area of your square, rectangular, or other complex-shaped ground grid into an equivalent area circle.

For this example, we will use a 30 m × 30 m ground grid (98.4 ft × 98.4 ft) with cross conductors every 5 m (16.4 ft). We will use a 4/0 AWG bare copper conductor buried 0.25 m (0.82 ft) below grade, with 10-ft ground rods driven at every point where the conductors cross or meet. To convert this square grid into an equivalent circle, we first need to calculate the area. A 30 m × 30 m grid is simply multiplied to get 900 m², as follows:

$$30 \text{ m} \times 30 \text{ m grid } (98.4 \text{ ft} \times 98.4 \text{ ft}) = 900 \text{ m}^2 \ (9682.6 \text{ ft}^2)$$

Next we need to plug 900 into the formula for the area of a circle. The area of a circle is:

$$A = \pi r^2$$

So, if we plug the numbers into the formula we get: $900 = 3.14 \times r^2$

$$900 \text{ m}^2 = \pi r^2$$

Next, we divide by π on both sides:

$$\frac{900\,\text{m}^2}{\pi} = \frac{\pi}{\pi} r^2$$

Since $900/3.14 = 286.48$, we get:

$$286.48\,\text{m}^2 = r^2$$

Next we square root both sides:

$$\sqrt{286.48\,\text{m}^2} = \sqrt{r^2}$$

We now have the equivalent radius of 16.93 m.

$$r = 16.93\,\text{m}$$

So, our 30 m × 30 m grounding/earthing grid has an equivalent area of a circle with a radius of 16.93 m.

Step 3: Now we can finally calculate the estimated grounding/earthing grid resistance by using the following simple formula:

$$R = \frac{\rho}{4r}$$

where R = the calculated resistance of the electrode in Ω, ρ = the resistivity of the bottom soil layer in ohm-metres, and r = equivalent grid radius in metres, from Step 2.

If we plug our numbers into this formula, we get the following:

$$R = \frac{17}{4 \times 16.93}$$

Since $4 \times 16.93 = 67.72$, we simply take 17 and divide it by 67.72, which will mean that:

$$R = 0.25$$

So the estimated DC RTG of our 30 m × 30 m grounding/earthing grid is 0.25 Ω.

In BS 7430:2011+A1:2015 Chapter 9.5.6 Mesh, this equation is given for calculating a mesh or grid. This equation uses diameter instead of radius, which results in the following equation:

$$R_m = 0.443\frac{\rho}{\sqrt{A}} + \frac{\rho}{L}$$

where R_m = the calculated resistance of the mesh or grid electrode in Ω, ρ = the resistivity of the soil in ohm-metres, A = the actual area covered by the mesh or grid in square metres (m²), and L = the total length of strip used in the mesh in metres (m).

Simplified Impedance Calculations for Ground grids.   To get a rough estimate of what the AC impedance to ground of your complex grounding/earthing electrode system (ground grid) will be, even in multilayer soils, the following simple formula is one way to get it. This formula will work for nearly any size ground grid of any shape, as long as the entire grounding/earthing grid has similar soil conditions.

For this formula, the depth of the grid must be buried less than 0.25 m (0.82 ft) in depth. We must also know the total length of buried horizontal conductor. For our 30 m × 30 m (98.4 ft × 98.4 ft) example ground grid that we have been using throughout this section, we have placed cross conductors at 5-m (16.4-ft) intervals. This means that there our example ground grid has fourteen 30-m long conductors, totalling 420 linear metres (1378 ft) of buried horizontal ground conductors. We have also previously calculated the equivalent radius of our ground grid to be 16.93 m. Here is the formula:

$$Z_{sg} = \rho \left(\frac{1}{4\pi} + \frac{1}{L} \right)$$

where Z_{sg} = the calculated impedance of the grid (system ground) in Ω, ρ = the bottom-layer resistivity of the soil in ohm-metres, L = total length of buried horizontal ground conductors in metres, and r = equivalent radius of the electrode system in metres.

If we plug these numbers into the formula we get the following (using 17-Ωm uniform soil):

$$Z_{sg} = 17 \left(\frac{1}{4 \times 16.93} + \frac{1}{420} \right)$$

If we solve the basics we get:

$$Z_{sg} = 17 \left(\frac{1}{67.72} + .00238 \right)$$

If we solve the division we get:

$$Z_{sg} = 17(.01477 + .00238)$$

This will give us:

$$Z_{sg} = 17 \times 0.1715$$

Now the final estimated impedance of a 30 m × 30 m ground grid buried less than 0.25 m below grade is:

$$Z_{sg} = 0.29155$$

An impedance of 0.29 Ω lines up much favourably with what a computer would calculate using 17-Ωm uniform soil. However, as you will next see this formula does not necessarily work so well when using different soil conditions.

Different Ground Grid Calculations Give Different Results. So far we have talked about at least five different ways to calculate a grounding/earthing electrode system:

1. Hand calculates the DC resistance of each individual piece of your electrode system in uniform soil, and then piece the individual results together (not shown in the chart below).
2. Estimate the DC resistance of the overall electrode system in uniform soil using an area-equivalent radius (circle).
3. Estimate the AC impedance of the overall electrode system in uniform soil using an area-equivalent radius (circle) and the total length of horizontal conductors.
4. Use a computer to calculate the impedance of the electrode system in uniform soil.
5. Use a computer to calculate the impedance of the electrode system in complex soil—this is the most accurate method and will provide estimates that are highly accurate when compared against actual field measurements, such as the three-point fall-of-potential method.

The chart below shows the same 30 m × 30 m ground grid compared to 2, 3, 4, and 5 above. The calculations are using uniform 17-Ωm soil and the complex soil is using the complete soil model as shown in the Soil Model section.

	Simplified Resistance Formula w/ Uniform 17-Ωm Soil	Simplified Impedance Formula w/ Uniform 17-Ωm Soil	Computer-Calculated Impedance w/ Uniform 17-Ωm Soil	*Computer-Calculated Impedance w/ Complex Soil (*most accurate)
30 m × 30 m Grounding/ Earthing Grid	0.25 Ω	0.29 Ω	0.24 Ω	3.91 Ω*

*An actual field measurement will return a result closest to this estimate.

As can be seen in the above spreadsheet, the estimated grounding grid resistance and impedance are dramatically different from the most accurate computer-calculated estimate in complex soils of 3.91 Ω. However, we do see that the estimated impedance formula does line up fairly closely to the computer estimate with uniform soil. However, this accuracy does not carry through when we look at other soil models, as can be seen in the chart below.

It is important to show what happens when we use some common sample two-layer (three with air) soil models. The following high-over-low versus the low-over-high soil models have been used as case study examples for many white papers, as it generally will represent a good spread the two-layer soil models we can expect to find in the field. The following chart shows even more dramatic differences between the computer-calculated results (that have been validated to come closest to actual field measurements) and the estimates from the simplified formulas:

	Simplified Resistance Formula	Simplified Impedance Formula	*Computer-Calculated Result
Low-over-high soil 100 Ωm for 10 ft over 1000 Ωm	14.77 Ω	17.15 Ω	6.23 Ω
High-over-low soil 1000 Ωm for 10 ft over 100 Ωm	1.48 Ω	1.715 Ω	3.15 Ω

*An actual field measurement will return a result closest to this estimate.

The data presented in the chart above used the bottom soil resistivity as required in the formulas, and provide us with varying results. As can be seen, the simplified formulas tend to provide overly low-estimated resistance/impedance-to-ground numbers when dealing with lower-resistivity soil, and overly high-estimated resistance/impedance-to-ground numbers when dealing with higher-resistivity soils.

These vast differences will become hugely important when we start calculating the GPR of the ground grid in the next section. Again, the use of computer modelling programs designed for such calculations is highly encouraged, and in most countries is legally mandated when dealing with grounding systems that provide human safety in high-voltage environments (such as 29 CFR 1910.269).

Hemisphere Radius.   The hemisphere radius calculation is a less common formula that is used by some telecommunications engineers for estimating the resistance of remote ground references when estimating 300-V lines for circuit protection. This very rough calculation uses the bottom soil layer of a soil model. Please see *McGraw-Hill's Standard Handbook of Electrical Engineers* or McGraw-Hill's *National Electrical Code 2014 Earthing & Grounding Handbook* for more information on 300-V lines.

$$R = \frac{\rho}{2\pi r}$$

where R = the calculated resistance of the electrode in Ω, ρ = the resistivity of the bottom layer of soil in ohm-metres, ρ = 3.142, and r = radius in metres.

CALCULATING AN ESTIMATED GROUND POTENTIAL RISE OF AN ELECTRODE SYSTEM (GROUND GRID)

It is important to understand that when electrical energy enters your grounding/earthing electrode system, whether from steady-state normal conditions or from an electrical fault, the electrode system is only one component of an electrical circuit, and is essence "in series" with the earth. Since the amount of current entering the earth is given, and the resistance/impedance to ground of our electrode system is fixed, the last variable left to be known is voltage. When a given amount of electrical energy (current) enters our grounding/earthing electrode system, the resultant voltage is known as the ground potential rise or GPR.

Note: When dealing with human safety, such as step and touch voltage hazards, we are primarily concerned with the differences in voltages that can occur within our below grade grounding/earthing electrode system (step voltages) and the differences in voltage that will occur between the facilities above-grade metallic objects (touch voltages), both the voltage differences between the individual metallic objects, and the differences in voltage between the metallic objects and the earth.

Other concerns that come into play with human safety include X/R ratios, zero-sequence impedances, DC offset currents, fault clearing time, tolerable body limits, effects of crushed rock, effects of nearby metallic objects (overhead conductors, buried metal pipes, cables, etc.), insulating gas type electrical systems, current division, conductor stress, permeability and permittivity of materials, time domain, circuit ladder networks, and many other engineering factors all relate to a proper human safety analysis. These human safety factors require a computer modelling program to be properly analyzed and will not be discussed in this section.

Keep in mind, if you have a facility that deals with voltages greater than 1000 V, or uses delta-delta type power (ungrounded power), most countries, including America, Canada, Australia, New Zealand, the EU, and many more, have legal requirements mandating that a professional analysis of your facility be conducted by an electrical engineering firm specializing in step and touch voltage hazards (see 29 CFR 1910.269). Some countries, such as Canada, have requirements that certain types of facilities must have GPRs less than 5000 V, and must have additional grounding added until the GPR is under that value.

Calculating the GPR of a ground grid or electrode system is very simple and follows Ohm's law. In essence, the GPR (V) of a grounding/earthing electrode system is the product of:

- The impedance or resistance of the electrode system (Z)
- The net fault current flowing into the electrode system (I)

$$V = Z \times I$$

The net fault current is typically obtained from a short-circuit analysis of your facility and most engineers use the three-phase-to-ground fault as a worst case fault although a single line-to-ground fault is the more realistic fault scenario.

Below you will find the same example grounding/earthing electrode systems we have used throughout this section, in the same sample soil structures, all with a 1000-A fault applied to the grid. As the GPR is a product of the fault current and resistance/impedance of the grid, the results are pretty straightforward.

	Simplified Resistance Formula w/ Uniform 17-Ωm Soil	Simplified Impedance Formula w/ Uniform 17-Ωm Soil	Computer-Calculated Impedance w/ Uniform 17-Ωm Soil	Computer-Calculated Impedance w/ Complex Soil*
30 m × 30 m Grounding/ earthing grid	0.25 Ω	0.29 Ω	0.24 Ω	3.91 Ω
GPR with a 1000-A fault	1020 V	290 V	240 V	3905 V

*An actual field measurement will return a result closest to this estimate.

	Simplified Resistance Formula	Simplified Impedance Formula	Computer-Calculated Result*
Low-over-high soil 100 Ωm for 10 ft over 1000 Ωm	14.77 Ω = 14,770-V GPR	17.15 Ω = 17,150-V GPR	6.23 Ω = 6232-V GPR
High-over-low soil 1000 Ωm for 10 ft over 100 Ωm	1.48 Ω = 1480-V GPR	1.715 Ω = 1715-V GPR	3.15 Ω = 3147-V GPR

*An actual field measurement will return a result closest to this estimate.

ELECTRODE CALCULATIONS CONCLUSION

Presented in the section above have been a number of simple formulas that can be used for calculating the RTG of simple grounding electrodes in homogeneous (uniform) soil conditions for DC systems. Unfortunately, few if any locations around the world actually have homogeneous soil. In fact, most grounding/earthing systems will be installed in soils are better modelled using three to five soil layers for calculating the electrode system.

Additionally, when dealing with high-frequency AC events such as lightning or noise, these events require far more complex equations that take into account the host of phenomena that occur in such electrical systems. The effects of impedance, permeability, permittivity, capacitance, conductance, inductance, leakage currents, electric fields, magnetic fields, interface layers, wave propagation, and many other factors must be taken into account. Unfortunately, these factors are simply too complex for simple hand calculations. As such, only a single impedance formula, which did not calculate the effects of material properties for use in homogenous soils, was presented as an example to show the beginnings of the complexities involved in accurately calculating grounding/earthing systems.

It should be clear that the formulas presented in this section should only be used for generalizations and estimations. Grounding/earthing systems that must meet specific engineering specifications should be designed using appropriate engineering software.

COMPUTER MODELLING SOFTWARE

All of the computer calculations used in this section were conducted using the RESAP (Soil Resistivity Analysis) and MALZ (Frequency Domain Grounding/Earthing Analysis) modules of the CDEGS Integrated Software for Power System Grounding/Earthing, Electromagnetic Fields and Electromagnetic Interference program (Application Version 14.3.20.3 and SES-Tech Installation Version 14.3.115), which is written and produced by Safe Engineering Services & technologies ltd., in Quebec, Canada (http://www.sestech.com).

The CDEGS software program is used by the IEEE, NFPA, ANSI, and many other standards and regulatory agencies to develop their requirements used the world over. It is currently the single most validated engineering software in the electrical grounding/earthing field with over 30 years of extensive scientific validation. These validations have utilized field tests and comparisons with analytical or published research results, which have been conducted by SES as well as other independent researchers, and are documented in hundreds of technical papers published in the most reputed international journals.

Chapter Nine

BELOW-GRADE EARTHING

Many Codes divide earthing into two distinct areas: equipment earthing and system earthing. Equipment earthing is the process of connecting above-grade equipment to the earth. In other words, how to properly bond earthing wires to equipment and route them through conduits, circuit-breaker boxes, and so on. System earthing is the process of intentionally making an electrical connection to the earth itself. This is the actual connection of metal to soil, and the minimum standards by which this connection is made. This process is often referred to as "earthing."

The goal of this chapter is to provide a basic knowledge of system earthing and "earthing" in an easy-to-read and understandable manner. We will not discuss above-grade wiring issues, except where needed. The topics that will be covered are system earthing, the benefits and features of the available earthing electrodes, the earth potential rise (EPR) hazards of high-current discharges, and the effects lightning strikes will have on an earthing system. We will also introduce the principles of proper soil testing, resistance-to-earth testing, proper test-well installation, and meter selection.

Both equipment earthing and system earthing are becoming more important as technology rapidly advances. Many of the latest and most advanced systems have stringent earthing requirements. Understanding the available electrical data through proper earth testing enables the electrical engineer to manage earthing systems that will meet specified earthing criteria.

Our goal is to provide the basic knowledge needed to understand and make the right choices when it comes to electrical earthing. Remember, "To protect what's above the earth you need to know what's in the earth."

INTRODUCTION

In the last few decades, much has been learned about the interaction between the earthing electrode and the earth, which is a three-dimensional electrical circuit.

Ultimately, it is the soil resistivity (and spatial variations thereof) that determines system design and performance. There are a number of different earthing electrodes in use today. They are the standard driven rod, advanced driven rod, earthing plate, concrete-encased electrode (sometimes called a *Ufer earth*), water pipes, and the electrolytic electrode.

ZONE (OR SPHERE) OF INFLUENCE

An important concept as to how efficiently earthing electrodes discharge electrons into the earth is called *the zone of influence*, which is sometimes referred to as the *sphere of influence* (see Fig. 9.1). The zone of influence is the volume of soil throughout which the electrical potential rises to more than a small percentage of the potential rise of the earth electrode, when that electrode discharges current into the soil. The greater the volume compared with the volume of the electrode, the more efficient the electrode. Elongated electrodes, such as earth rods, are the most efficient. The surface area of the electrode determines the ampacity of the device, but does not affect the zone of influence. Case in point, the greater the surface area, the greater the contact with the soil and the more electrical energy that can be discharged per unit of time.

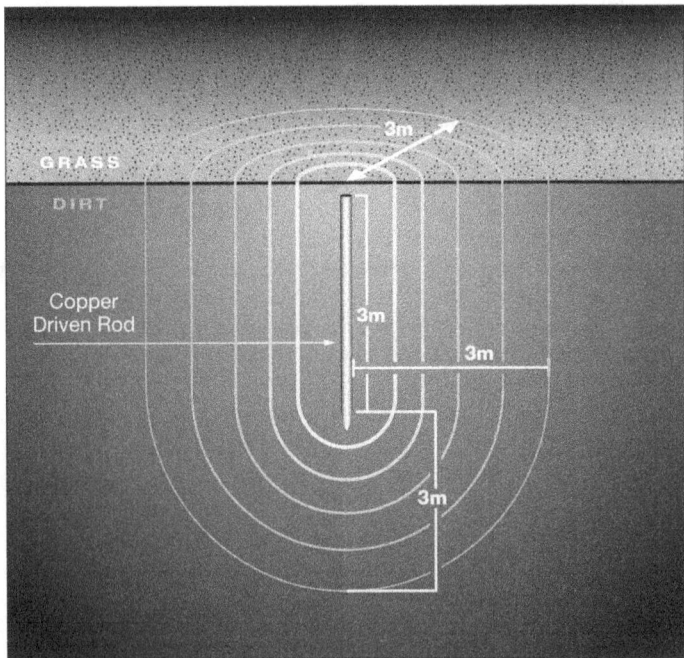

Figure 9.1 Sphere of influence of an earth electrode.

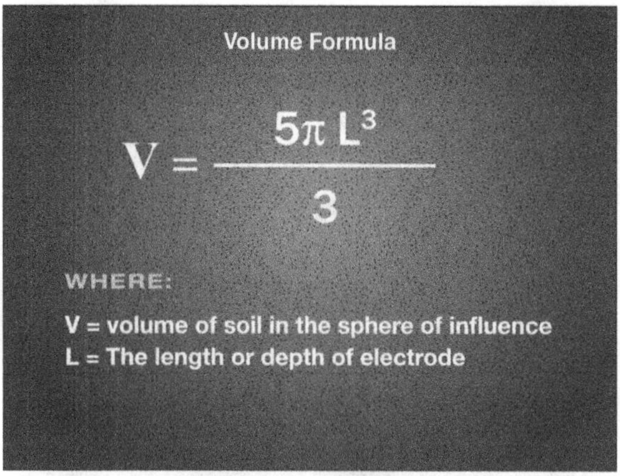

Figure 9.2 Volume of the sphere of influence.

The formula for calculating the volume of soil is shown in Fig. 9.2. A simpler version is used when the above formula is modified by rounding π (pi) down to 3 and cross-cancelling to get the formula:

$$V = 5L^3$$

Thus, a single 3-m driven rod will utilize 141.6 m^3 of soil, where as a single 2.4-m rod will utilize about half the soil at 72.5 m^3. Going from 2.4-m to 3-m earth rod can provide a significant reduction in the resistance to earth (RTE) as the sphere of influence will be nearly doubled, given that the soil resistivity does not increase with depth.

EARTHING ELECTRODES

Earthing is the process of electrically connecting any metallic object to the earth by way of an earth electrode system. Many Codes require that the earthing electrodes be tested to ensure that they are under 25 Ω resistance to earth (earth). It is important to know that aluminium electrodes are not allowed for use in earthing (earthing) as aluminium will rapidly corrode when buried.

Driven Rod. The standard driven rod or copper-clad rod consists of an 8–10 ft length of steel with a 5–10 mil coating of copper (see Fig. 9.3). This is by far the most common earthing device used in the field today. The driven rod has been in use since the earliest days of electricity, with a history dating as far back as Benjamin Franklin.

Driven rods are relatively inexpensive to purchase; however, ease of installation is dependent upon the type of soil and terrain where the rod is to be installed. The steel used in the manufacture of a standard driven rod tends to be relatively soft. Mushrooming can occur on both the tip of the rod, as it encounters rocks on its way down, and the end where force is being applied to drive the rod through the

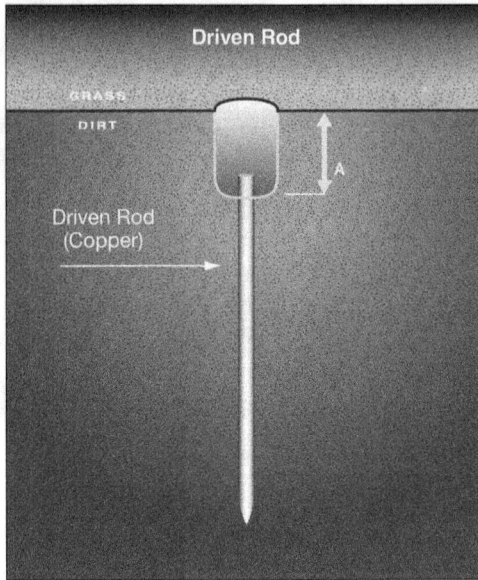

Figure 9.3 Copper-clad driven earthing rod.

earth. Driving these rods into the earth can be extremely labour intensive, when rocky terrain creates problems as the tips of the rods will continue to mushroom. Often, these rods will hit a rock and actually turn back around on themselves and pop back up a few feet away from the installation point.

Because driven rods range in length from 2.4 to 3 m, a ladder is often required to reach the top of the rod, which can become a safety issue. Many falls have resulted from personnel trying to literally "whack" these rods into the earth while hanging from a ladder many feet in the air.

Many standards require that driven rods be a minimum of 2.4 m in length and that 2.4 m of length must be in direct contact with the soil. To comply with this requirement, the installer will typically use a shovel to dig down into the earth by about 0.46 m before a driven rod is installed, although the most common rods used by commercial and industrial contractors today are rods 3 m in length, which negates the need for the extra installation process. This can save time as well as meet with the many industrial specifications that also require this length as a minimum.

A common misconception is that the copper coating on a standard driven rod has been applied for electrical reasons. While copper is certainly a conductive material, its real purpose on the rod is to provide corrosion protection for the steel underneath. Many corrosion problems can occur because copper is not always the best choice in corrosion protection. It should be noted that galvanized driven rods have been developed to address the corrosion concerns that copper presents, and in many cases are a better choice for prolonging the life of the earthing rod and earthing systems. Generally speaking, galvanized rods are a better choice in all but high-salt environments.

An additional drawback of the copper-clad driven rod is that copper and steel are two dissimilar metals. When an electrical current is imposed on the rod, electrolysis will occur, also the act of driving the rod into the soil can further damage the copper cladding, allowing corrosive elements in the soil to attack the bared steel and decrease the life expectancy of the rod. Environment, aging, temperature, and moisture also easily affect driven rods, giving them a typical life expectancy of 5–15 years in good soil conditions. Driven rods also have a very small surface area, which is not always conducive to good contact with the soil. This is especially true in rocky soils, in which the rod will only make contact on the edges of the surrounding rock.

A good example of this is to imagine a driven rod surrounded by large marbles. Actual contact between the marbles and the driven rod will be very small. Because of this small surface contact with the surrounding soil, the rod will increase in RTE, lowering the conductance, and limiting its ability to handle high-current faults.

EXAMPLE ELECTRODE INSTALLATIONS FROM THE U.S. NATIONAL ELECTRICAL CODE (NEC)

The British system of standards really doesn't have a great number of details regarding the way earthing electrodes are to be installed. Due to the number of questions received over the years, it is clear that some of the methods used under the NEC would be valuable information for IEC users. The following section covers typical electrode installation methods under the NEC.

In order to qualify as a permitted earthing electrode, all rod and pipe electrodes must be at least 8 ft in length (or 2.44 m), installed below permanent moisture level, and in direct contact with the earth. This is why 3-m (10-ft.) earth rods are recommended when installing the earth rod with a portion of the rod above-grade level. When a 3-m (10-ft) earth rod is installed, you can have 0.15 m (6 in.) exposed above grade, 0.46 m (18 in.) in the permanent moisture level (commonly thought of as being 0.46 m deep), and the mandatory 8 ft of rod in contact with the earth (see Fig. 9.4).

Permanent moisture level is undefined within the NEC, but is commonly linked to either the frost line or the permanent wilting point of the site. Permanent moisture level is also often thought of as being either 0.46 m (1.5 ft) or 0.76 m (2.5 ft) in depth, due to various interpretations of several codes, regulations, and industrial standards.

Earth rods (rod-type electrodes) must have a diameter of at least 15.87 mm (5/8 in.), and must be made of either zinc or copper-coated steel, or be solid stainless steel. Of course, there are a few other listed electrode systems that fall in this category, and may of course be used.

Pipe-type electrodes must have a diameter of at least trade size ¾ in. (metric designator 21) and must be coated with a corrosion resistant metal such as galvanization. Copper pipes are considered to be corrosion resistant.

When installing an earth rod below grade, it is best to have the top of the rod below the permanent moisture level of 0.46 m (18 in.), in order to protect the rod from corrosion and to maximize the benefits of the sphere of influence of the rod. The connection to the earth rod must be made with connections suitable for direct

MECHANICAL CLAMP
IN THE AIR
(ABOVE GROUND)

.5 FEET
(6 INCHES)

2 FEET

1.5 FEET
(10 INCHES)

PERMANENT
MOISTURE
LEVEL - 18in.

8-FOOT
MINIMUM
CONTACT

6-FEET MIN./
20-FEET IDEAL

10-FOOT
ROD

(10-FOOT ROD
REQUIRED)

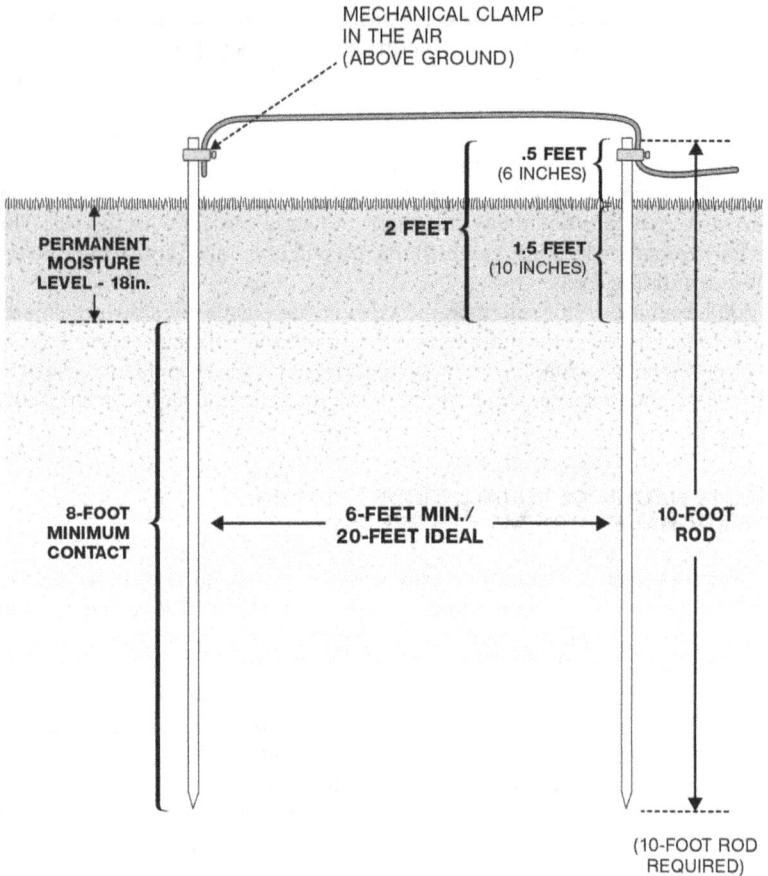

Figure 9.4 Earth rod installation with above-grade connection.

burial. Remember that 3-m rods have twice the sphere of influence as a 2.4-m rod, so the longer earth rod is a superior choice (see Fig. 9.5).

Sometimes earth rods will not drive all the way into the earth to the desired depth. This is called earth rod refusal, and the NEC has a specific order that is to be used to compensate. Obviously, the ideal method for earth rod installation is to install the rod straight down into the earth. If earth rod refusal happens, the next choice is to install the rod at a 45° angle, sometimes called *crow's foot*. The last method is to install the rod in a trench at least 0.76 m (2.5 ft.) below-grade level.

The NEC tells us that if you have a single ground rod, pipe, or plate electrode, you must also have an additional qualified electrode, that is bonded to the rod, pipe, or plate electrode, or is bonded to the grounding electrode conductor, to the neutral conductor at the service entrance, to non-flexible grounded service raceway, or to any grounded service enclosure.

CLAMPS SUITABLE FOR
DIRECT BURIAL OR
EXOTHERMIC WELD

PERMANENT
MOISTURE
LEVEL - 18in.

6-FEET MIN./
20-FEET IDEAL

8- OR 10-
FOOT ROD

(10-FOOT ROD
PREFERRED)

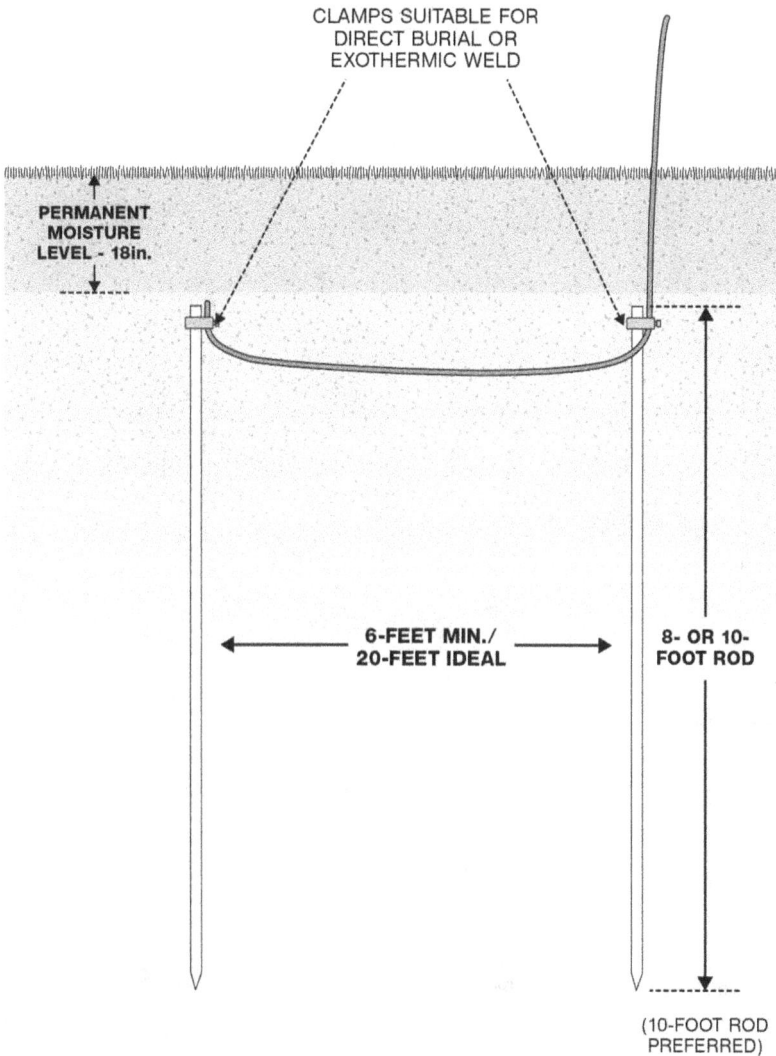

Figure 9.5 Earth rod installation with below-grade connection.

The NEC also states that your 3-m earth rods (or pipe/plate electrodes) should be spaced 6 m (2x the length of the electrode) apart. But if room does not exist, you may bring them in as close as 2 m from each other (see Fig. 9.6).

The electrodes should be spaced at twice the distance of the longest electrode. For instance, if you have 2.4-m (8-ft) earth rods (which you should not use for reasons mentioned earlier), you should install them 4.8 m (16 ft) apart. If you have 3-m (10-ft) earth rods (which you should always use), you should install them 6 m (20 ft) apart. This is because earth rods are more effective if they do not have overlapping sphere of influence.

GROUNDING ELECTRODE
CONDUCTORS

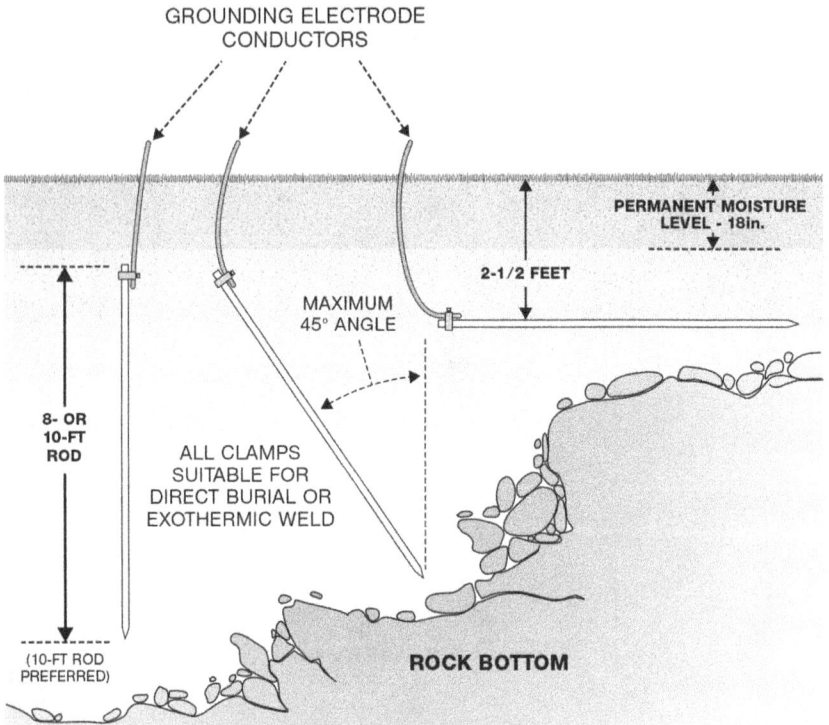

PERMANENT MOISTURE
LEVEL - 18in.

2-1/2 FEET

MAXIMUM
45° ANGLE

8- OR
10-FT
ROD

ALL CLAMPS
SUITABLE FOR
DIRECT BURIAL OR
EXOTHERMIC WELD

(10-FT ROD
PREFERRED)

ROCK BOTTOM

Figure 9.6 Three options for earth rod installation.

Advanced Driven Rods. Advanced driven rods are specially engineered variations of the standard driven rod, with several key improvements. Because they
present lower physical resistance, advanced rods can now go into terrain where
only large drill rigs could install before and can quickly be installed in less
demanding environments. The modular design of these rods can reduce safety-
related accidents during installation. Larger surface areas can improve electrical
conductance between the soil and the electrode.

Of particular interest is that advanced driven rods can easily be installed to
depths of 6 m or more, depending upon soil conditions (see Fig. 9.7).

Advanced driven rods are typically driven into the earth with a standard drill
hammer. This automation dramatically reduces the time required for installation.
The tip of an advanced driven rod is typically made of carbide and works in a

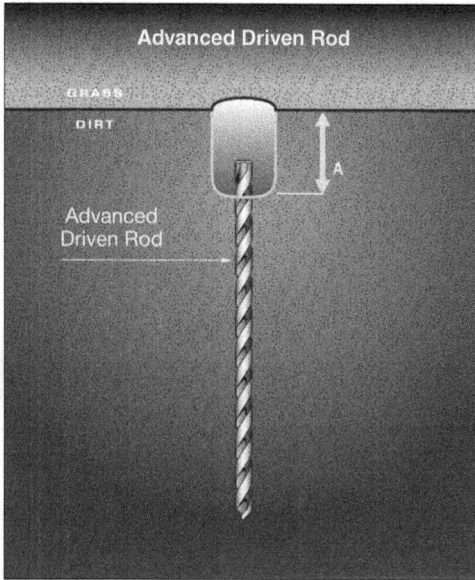

Figure 9.7 Advanced driven earthing rod.

similar manner to a masonry drill bit, allowing the rod to bore through rock with relative ease. Advanced driven rods are modular in nature and are designed in 5 ft lengths. They have permanent and irreversible connections, which enable an operator to install them safely, while standing on the earth. Typically, a shovel is used to dig down into the earth 0.46 m before the advanced driven rod is installed. The advanced driven rod falls into the same category as a driven rod and satisfies the same codes and regulations.

In the extreme northern and southern climates of the planet, frost heave is a major concern. As frost sets in every winter, unsecured objects buried in the earth tend to be pushed up and out of the earth. Driven earthing rods are particularly susceptible. Anchor plates are sometimes welded to some buried portion of the rods to prevent them from being pushed up and out of the earth by frost heave. This however requires that a hole be augured into the earth in order to get the anchor plate into the earth, which can dramatically increase installation costs. Advanced driven rods do not suffer from frost heave issues and can be installed easily in extreme climes.

Earthing Plates. Earthing plates are typically thin copper plates buried in direct contact with the earth. Many Codes require that earth plates have at least 0.186 m^2 of surface area exposed to the surrounding soil. Ferrous materials must be at least 5 mm thick, whereas non-ferrous materials (copper) need only be 15 mm thick. Earthing plates are typically placed under poles or supplementing buried earth rings.

As shown in "A" on Fig. 9.8, earthing plates should be buried at least 30 in. Below-grade level. While the surface area of earthing plates is greatly increased over that of a driven rod, the zone of influence is relatively small as shown in "B" on Fig. 9.8 The zone of influence of an earthing plate can be as small as 17 in.

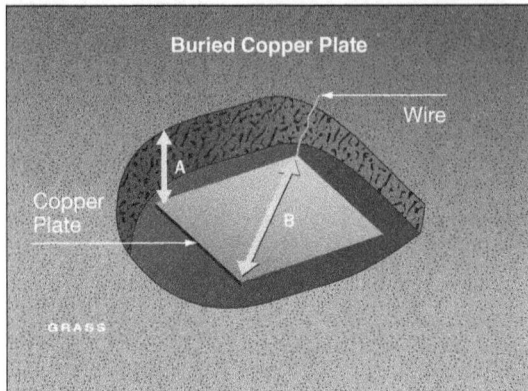

Figure 9.8 Buried copper plate.

This ultrasmall zone of influence typically causes earthing plates to have a higher resistance reading than other electrodes of the same mass. Similar environmental conditions, which lead to the failure of the driven rod, also plague the earthing plate, such as corrosion, aging, temperature, and moisture.

Ufer Earth or Concrete-Encased Electrodes. Originally, Ufer earths were copper electrodes encased in the concrete surrounding ammunition bunkers. In today's terminology, Ufer earths consist of any concrete-encased electrode, such as the rebar in a building foundation, when used for earthing, or a wire or wire mesh encased in concrete.

Concrete-Encased Electrode. Many Codes require that concrete-encased electrodes use a minimum 5.88-mm copper wire at least 6 m in length and encased in at least 2 in. of concrete. The advantages of concrete-encased electrodes are that they dramatically increase the surface area and degree of contact with the surrounding soil. However, the zone of influence is not increased; therefore, the RTE is typically only slightly lower than the wire would be without the concrete (see Fig. 9.9).

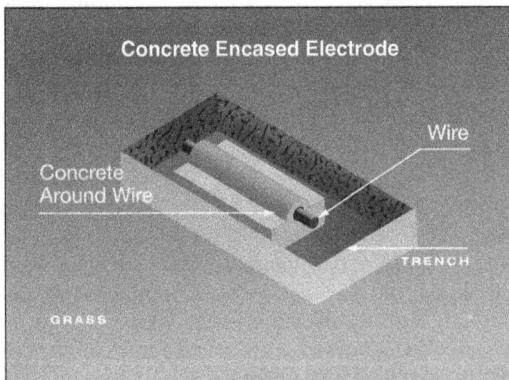

Figure 9.9 Concrete-encased electrode.

Concrete-encased electrodes also have some significant disadvantages. When an electrical fault occurs, the electric current must flow out of the conductor and through the concrete to get to the earth. Concrete, by nature, retains a lot of water that rises in temperature as the electricity flows through the concrete. If the concrete-encased electrode is not sufficient to handle the total current, the boiling point of the water may be reached, resulting in an explosive conversion of water into steam. Many concrete-encased electrodes have been destroyed, after receiving relatively small electrical faults. Once the concrete cracks apart and falls away from the conductor, the concrete pieces act as a shield preventing the copper wire from contacting the surrounding soil, resulting in a dramatic increase in the RTE of the electrode.

There are many new products available on the market designed to improve concrete-encased electrodes. The most common are modified concrete products that incorporate conductive materials into the cement mix, usually carbon. The advantage of these products is that they are fairly effective in reducing the resistivity of the concrete, thus lowering the RTE of the electrode encased. The most significant improvement of these new products is in reducing heat buildup in the concrete during fault conditions, which can lower the chances that steam will destroy the concrete-encased electrode. However, some disadvantages are still evident. Again, these products do not increase the zone-of-influence and as such, the RTE of the concrete-encased electrode is only slightly better than what a bare copper wire or driven rod would be in the earth. Also, a primary concern regarding enhanced earthing concretes is the use of carbon in the mix. Carbon and copper are of different nobilities and will sacrificially corrode each other over time. Many of these products claim to have buffer materials, designed to reduce the accelerated corrosion of the copper, caused by the addition of carbon into the mix. However, few independent long-term studies are being conducted to test these claims.

Ufer Earth or Building Foundations.   Ufer Earths or building foundations may be used provided that the concrete is in direct contact with the earth (no plastic moisture barriers), that rebar is at least 0.500 in. in diameter and that there is a direct metallic connection from the service earth to the rebar buried inside the concrete (see Fig. 9.10).

This concept is based on the conductivity of the concrete and the large surface area, which will usually provide an earthing system, which can handle very high-current loads. The primary drawback occurs during fault conditions, if the fault current is too great compared with the area of the rebar system, when moisture in the concrete superheats and rapidly expands, cracking the surrounding concrete, threatening the integrity of the building foundation. Another important drawback to the Ufer earth is that they are not testable under normal circumstances, as isolating the concrete slab in order to properly perform RTE testing, is nearly impossible.

The metal frame of a building may also be used as an earthing point, provided that the building foundation meets the above requirements, and is commonly used in high-rise buildings. It should be noted that many owners of these high-rise buildings are banning this practice and insisting that tenants run earth wires all the way back to the secondary service locations on each floor. The owners will already have run earth wires from the secondary services back to the primary service locations and installed dedicated earthing systems at these service locations.

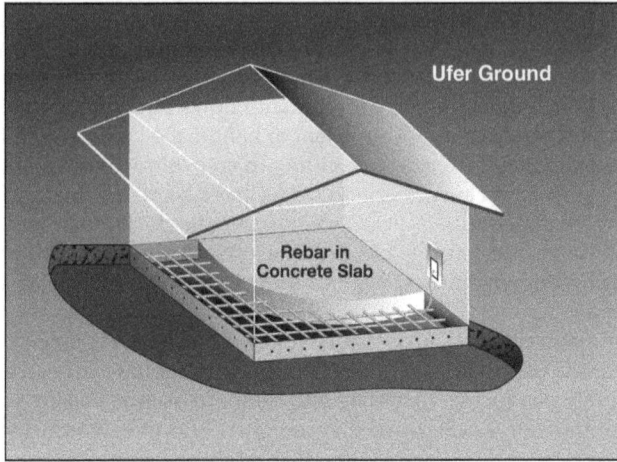

Figure 9.10 Building foundation or Ufer.

The goal is to avoid the flow of stray currents, which can interfere with the operation of sensitive electronic equipment.

Example Concrete Connection from the U.S. National Electrical Code (NEC). In regards to steel rebar in concrete, the NEC requires that the junction point at where a conductor enters the concrete must have a mechanical stress relief to prevent the shearing of the conductor. Typically, a piece of PVC pipe with silicone caulking is used to accomplish this requirement (see Fig. 9.11). However, a number of commercially available products provide superior connection methods that should be considered.

Note: There are a number of products designed to reduce the resistance of concrete and theoretically enhance the ability of the concrete to conduct electricity. These products should be avoided as they tend to utilize carbon-based materials, which will accelerate the rate of corrosion for copper.

Water Pipes. Water pipes have been used extensively over time as an earthing electrode. Water pipe connections are not testable and are unreliable due to the use of tar coatings and plastic fittings. City water departments have begun to specifically install plastic insulators in the pipelines, to prevent the flow of current and reduce the corrosive effects of electrolysis. Many Codes require that at least one additional electrode be installed, when using water pipes as an electrode. There are several additional requirements including:

- 3 m of the water pipe is in direct contact with the earth.
- Joints must be electrically continuous.
- Water meters may not be relied upon for the earthing path.
- Bonding jumpers must be used around any insulating joints, pipe, or meters.
- Primary connection to the water pipe must be on the street side of the water meter.
- Primary connection to the water pipe shall be within 5 ft of the point of entrance to the building.

GROUNDING ELECTRODE
CONDUCTOR

4 AWG OR LARGER
BARE COPPER CONDUCTOR
OR STEEL REINFORCING BAR
OR ROD NOT LESS THAN
1/2-IN. IN DIAMETER,
AT LEAST 20-FT LONG

NON-METALLIC
PROTECTIVE SLEEVE

CONNECTION LISTED
FOR THE PURPOSE

20 FT. OR MORE

2 IN
MIN.

FOUNDATION IN
DIRECT CONTACT
WITH THE EARTH

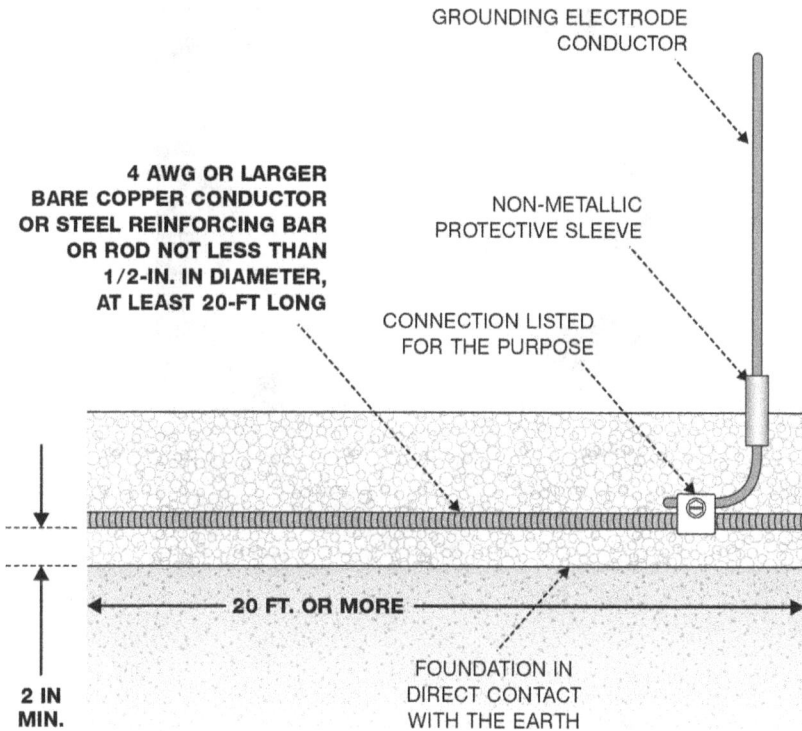

Figure 9.11 Mandatory earthing connection to steel rebar in concrete.

Many Codes require that water pipes be bonded to earth, even if water pipes are not used as an earthing electrode.

Electrolytic Electrode.   The electrolytic electrode was specifically engineered to eliminate many of the drawbacks found in other types of earthing electrodes. The electrolytic electrode consists of a hollow copper shaft, filled with salts and desiccants whose hygroscopic nature draws moisture from the air. The moisture mixes with the salts to form an electrolytic solution, which continuously seeps into the surrounding backfill material, keeping it moist and high in ionic content. The electrolytic electrode is installed into an augured hole and is typically backfilled with a conductive material, such as bentonite clay (see Fig. 9.12). The electrolytic solution and the backfill material work together to provide a solid connection between the electrode and the surrounding soil, that is free from the effects of temperature, environment, and corrosion. The electrolytic electrode is the only earthing electrode that improves with age. All other electrode types will have a rapidly increasing RTE as the seasons change and the years pass. The drawbacks to these electrodes are the cost of installation and the cost of the electrode itself.

Various backfill products are available in the market place; the primary concern should be if the product protects the electrode from corrosion and improves its conductivity. Carbon-based products should be avoided as they will corrode the copper over time.

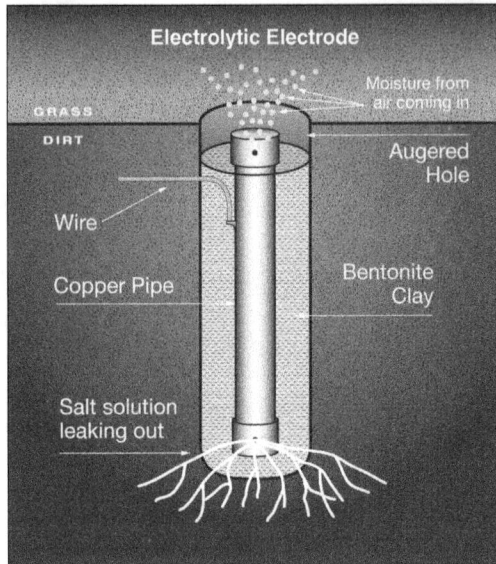

Figure 9.12 Electrolytic electrode.

There are generally two types of electrolytic electrodes that one can install, ones that use sodium chloride (table or rock salt), and those that use magnesium sulphate (Epsom salt). There are advantages and disadvantages for each type. The electrolytic electrodes that use sodium chloride have very long life spans (30–50 years) and as such are often sealed closed as there is no need to access the tube. The disadvantage is that very little salt actually enters the surrounding soil, so the time it takes to lower the RTE can be very long (years if not decades). The electrolytic electrodes that use magnesium sulphate come with an access cap at the top of the electrode, as the magnesium sulphate will rapidly dissolve away and out of the tube entering the surrounding soil, thus quickly lowering the RTE. The disadvantages of magnesium sulphate electrodes is that they require annual maintenance to refill the salts in the tube, and if the magnesium sulphate is exposed to high-heat, such as from a lightning strike, chemical reactions can occur resulting in some toxic substances. The Chemicals Hazard Information and Packaging for Supply (CHIP) or Material Safety Data Sheet (MSDS) sheet for magnesium sulphate should be consulted prior to use. Some earthing engineers have installed electrolytic electrodes with magnesium sulphate for the first few years of operation so as to rapidly lower the RTE, and then switched over to a sodium chloride and desiccant mix for the long life and low maintenance once the surrounding soil has been saturated with conductive materials.

Earth-Electrode Comparison Chart. The following chart (Table 9.1) compares the various types of electrodes versus some important characteristics that may prove helpful in selecting proper electrode usage.

Earthing Electrode Test Wells. The purpose of an earth test well is twofold: to be able to quickly and easily measure RTE of the earthing electrode and/or to measure point-to-point continuity (resistance) from one electrode to another.

Table 9.1: Earth Electrode Comparison Chart

	Driven Rod	Advanced Driven Rod	Earthing Plate	Concrete-Encased Electrode	Building Foundation	Water Pipe	Electrolytic Electrode
Resistance to earth (RTE)	Poor	Average	Poor	Average	Above average	Poor to excellent**	Excellent
Corrosion resistance	Poor	Good	Poor	Good*	Good*	Varies	Excellent
Increase in RTE in cold weather	Worsens	Slightly worsens	Worsens	Slightly worsens	Slightly worsens	Minimally Affected	Minimally affected
Increase in RTE over time	Worsens	Typically unaffected	Worsens	Typically unaffected	Typically unaffected	Typically unaffected	Improves
Electrode ampacity	Poor	Average	Average	Average*	Above average*	Poor to excellent**	Excellent
Installation cost	Average	Excellent	Below average	Below average	Average	Average	Poor
Life expectancy	Poor 5–10 years	Average 15–20 years	Poor 5–10 years	Average* 15–20 years	Above average* 20–30 years	Below average 10–15 years	Excellent 30–50 years

*High-current discharges can damage foundations when water in the concrete is rapidly converted into steam.
**When part of extensive, bare, metallic, electrically continuous water system.

The reason for this section is that most earthing test wells are improperly installed, preventing accurate RTE measurements of the electrode.

The first part, RTE, is a measurement of the total resistance to the flow of electricity that the earth is providing against the earth system under test. As the three-point fall-of-potential method is not usable for any active earthing system, only the clamp-on or induced frequency method can be used for testing the RTE of electrodes with a test well. This involves a hand-held meter with large jaws that are "clamped" around an earthing conductor under test, thus the need for a test well.

The second part, which is simple point-to-point continuity (resistance) measurements, can be done even at an improperly installed test well. Point-to-point testing from an earth test well back to a reference point can be very valuable for understanding the integrity and conductivity of an earthing system. The goal is to install earth test wells that are capable of allowing both measurements.

Improper Installation of an Earth Test. One of the most common ways earth test wells are improperly installed, as seen in the image labelled Fig. 9.13. In this scenario, the installer has simply provided physical access to the earth ring with an earth rod "T-welded" into the loop. This scenario allows us to clamp on to the earth loop conductor above the earth rod itself. However, the signal from the clamp-on earth resistance meter will merely travel around the copper conductor in a loop, and is never forced to travel through the earth itself. This test well is improperly installed and will not allow an accurate resistance-to-earth measurement.

In Fig. 9.14, we see the second most common way test wells are improperly installed, and this is with a twist or "loop" added to the earth conductor. As you can see, this scenario is actually the exact same thing as scenario 1, although quite possibly easier to clamp the meter around the conductor. The downside is that this twist in the conductor is a major violation of the rules listed in NFPA 780 Lightning Standards, and the Motorola R56 (and several IEEE standards) in regards to the self-induced coupling effects that high-current, short-duration faults can have

Figure 9.13 Incorrect test well.

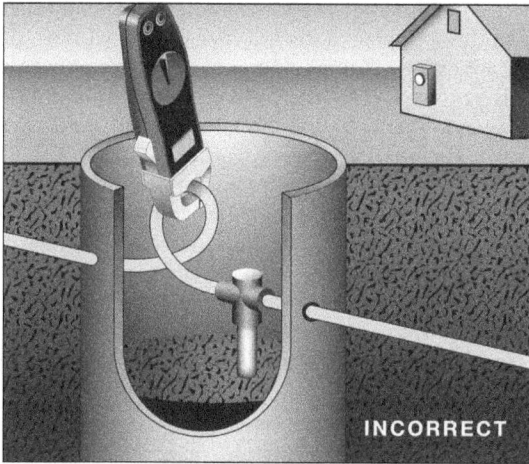

Figure 9.14 Incorrect test well with a loop.

on earth systems. When lightning strikes or a major short-circuit fault occurs, a large and powerful magnetic field will form as the current travels through the earth conductor. Any conductors with a "tight radius" are subject to burn open due to cross coupling of the magnetic fields. The bottom line is that this test well is improperly installed and not only is it at risk of critical failure under electrical stress, but it will not allow an accurate resistance-to-earth measurement test.

Proper Installation of an Earth Test Well.   In Fig. 9.15, we can see that installing an earthing electrode in a test well simply requires a short conductor extension or "pig-tail" to be added connecting the top of the electrode to the earth loop. While

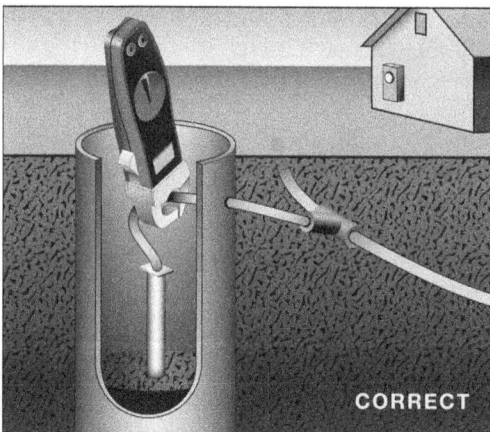

Figure 9.15 Correct test well.

this does require two welds instead of just a single exothermic weld, it is often less labour intensive as the conductor extension is easier to work with than the earth loop. In this setup, we can see that the clamp-on meter has easy access and that the injected signal will be forced down the earth rod and through the earth, enabling an accurate resistance-to-earth reading of that earth rod. And of course, accurate point-to-point testing can be conducted as well.

It is important to note that this test well scenario is only valid for testing a single earthing electrode at a time, within the bigger earthing system. It does not provide a RTE of your entire earth system. Based on the size and requirements of your earth system, it may not even be possible to design an earthing system that can be tested as a whole. This is why multiple earth test wells are often installed at key locations around your earth system; they provide a means of validating the earth system integrity at multiple locations for comparison over time.

SAMPLE PROTECTIVE BONDING REQUIREMENT FROM THE U.S. NATIONAL ELECTRICAL CODE (NEC)

Figure 9.16, *Key Articles of the Code*, shows all of the various earthing connections required under the NEC. This illustration is actually fairly shocking to many electrical engineers and electricians in NEC countries, as they are completely unaware that all of the earthing systems for these items are required to be bonded together.

Unlike BS 7671 Section 411.3.1.2, which provides a clear list of the metallic items within an installation that are to be bonded to the earthing system, the NEC fails to give a single comprehensive list and instead spreads the various requirements all over the code.

This illustration does of course apply to IEC codes (including BS 7671); they simply have different regulation numbers. In the case of BS 7671, we would simply list Section 411.3.1.2 (there are of course numerous other code regulation within BS 7671 that also deal with PE bonding) and be done with it.

While this illustration is fairly redundant for BS 7671 users, it is still a great visual and hopefully you will find it useful.

System Design and Planning.   An earthing design starts with a design specification and is often followed by a site analysis, collection of geological data, and soil resistivity of the area. Typically, the site engineer or equipment manufacturers specify a minimum RTE number. The NEC states that the RTE shall not exceed 25 Ω for a single electrode. However, high-technology manufacturers will often specify 3 or 5 Ω, depending upon the requirements of their equipment. For sensitive equipment and under extreme circumstances, a 1 ohm specification may sometimes be required. When designing an earth system, the difficulty and cost increase exponentially as the target RTE approaches the unobtainable goal of 0 Ω.

Data Collection.   Once the specifications have been established, data collection begins. Soil resistivity testing, geological surveys, and test borings should provide the basis for all earthing designs. Proper soil resistivity testing using the Wenner four-point method is recommended because of its accuracy. This method will be discussed later in this chapter. Additional data is always helpful and can be

KEY ARTICLES
OF THE CODE

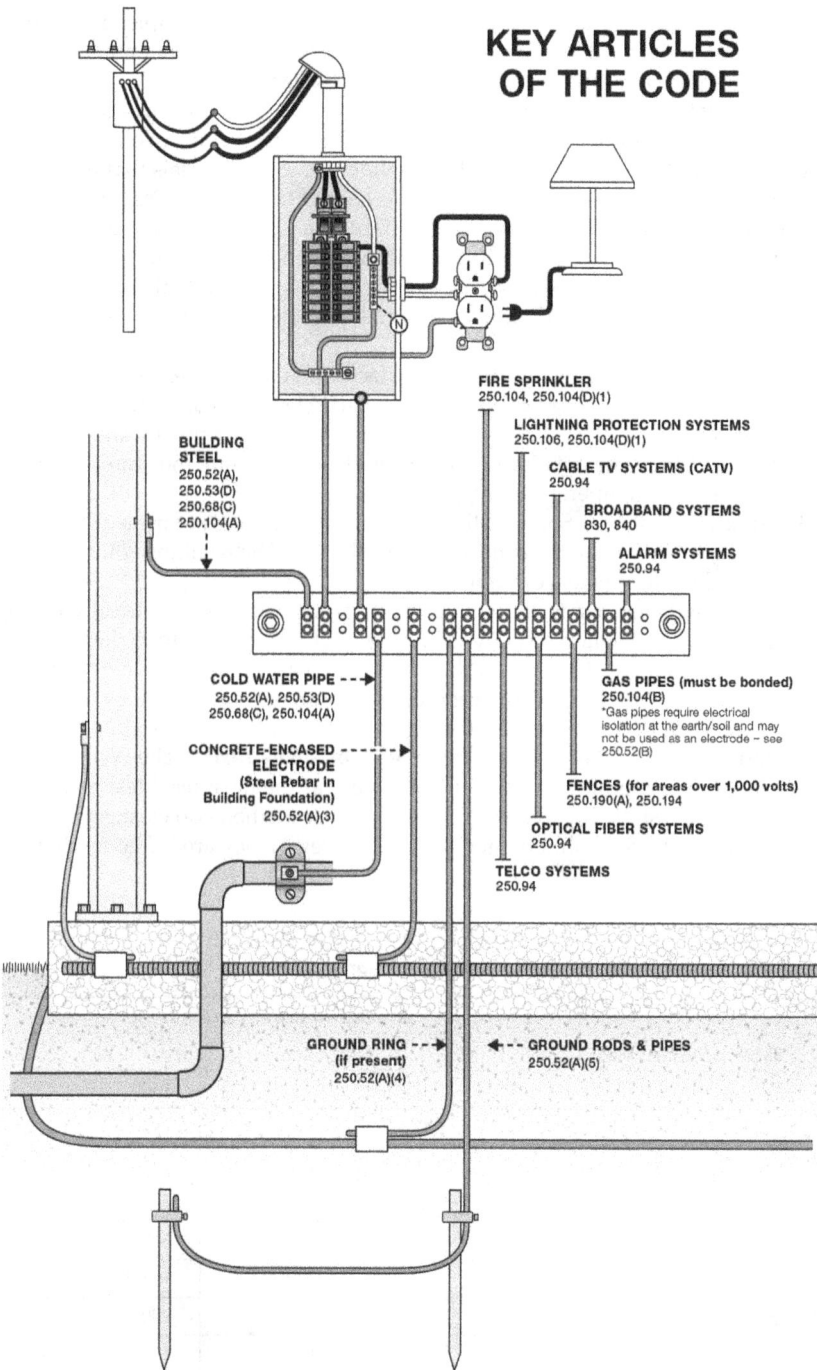

FIRE SPRINKLER
250.104, 250.104(D)(1)

LIGHTNING PROTECTION SYSTEMS
250.106, 250.104(D)(1)

BUILDING
STEEL
250.52(A),
250.53(D)
250.68(C)
250.104(A)

CABLE TV SYSTEMS (CATV)
250.94

BROADBAND SYSTEMS
830, 840

ALARM SYSTEMS
250.94

COLD WATER PIPE
250.52(A), 250.53(D)
250.68(C), 250.104(A)

GAS PIPES (must be bonded)
250.104(B)
*Gas pipes require electrical
isolation at the earth/soil and may
not be used as an electrode – see
250.52(B)

CONCRETE-ENCASED
ELECTRODE
(Steel Rebar in
Building Foundation)
250.52(A)(3)

FENCES (for areas over 1,000 volts)
250.190(A), 250.194

OPTICAL FIBER SYSTEMS
250.94

TELCO SYSTEMS
250.94

GROUND RING
(if present)
250.52(A)(4)

GROUND RODS & PIPES
250.52(A)(5)

Figure 9.16 NEC mandatory earthing connections.

collected from existing earth systems located at the site. For example, driven rods at the location can be tested using the three-point fall-of-potential method or an induced frequency test using a clamp-on earth resistance meter.

Data Analysis.   Rarely is soil uniform in resistivity from the surface down to depth equal to the whole earthing systems sphere of influence. Soil is typically organized into horizontal layers of largely homogenous materials that have been laid down through the ages, each with different electrical properties. The resistivity of these layers can be inferred with careful measurements and sophisticated computer modelling techniques. The results are often that the actual soil resistivities from layer to layer can vary by many orders of magnitude. Only by calculating the resistivity and depths of these layers (called a soil model) can you accurately design an earthing system.

Earthing Design.   Soil resistivity is the key factor that determines the resistance or performance of an earthing system. It is the starting point of any earthing design. As you can see in Tables 9.2 and 9.3, soil resistivity varies dramatically throughout the world and is heavily influenced by electrolyte content, moisture, minerals, compactness, and temperature.

Soil Resistivity Testing.   Soil resistivity testing is the process of measuring a volume of soil, to determine the conductivity of the soil. The resulting soil resistivity is expressed in ohm-meters or ohm-centimetres.

Soil resistivity testing is the single most critical factor in electrical earthing design. This is true when discussing simple electrical design, to dedicated low-resistance earthing systems, or to the far more complex issues involved in EPR studies. Good soil models are the basis of all earthing designs and they are developed from accurate soil resistivity testing.

Wenner Soil Resistivity Test and Other Soil Resistivity Tests.   The Wenner four-point method (sometimes called four-pin) is by far the most used test method to measure the resistivity of soil (see Fig. 9.17). Other methods do exist, such as the General method and Schlumberger method, however, they are infrequently used

Table 9.2: Surface Materials versus Resistivity Chart

Type of Surface Material	Average Resistivity in Ohm-Meters	
	Dry	Wet
Crusher granite w/fines	140×10^6	1300
Crusher granite w/fines 1.5"	4000	1200
Washed granite – pea gravel	40×10^6	5000
Washed granite 0.75"	2×10^6	10,000
Washed granite 1–2"	1.5×10^6 to 4.5×10^6	5000
Washed granite 24"	2.6×10^6 to 3×10^6	10,000
Washed limestone	7×10^6	2000 to 3000
Asphalt	2×10^6 to 30×10^6	10,000 to 6×10^6
Concrete	1×10^6 to 1×10^9	20 to 100

Table 9.3: Soil Types versus Resistivity Chart

Soil Types or Type of Earth	Average Resistivity in Ohm-Meters
Bentonite	2–10
Clay	20–1000
Wet organic soils	10–100
Moist organic soils	100–1000
Dry organic soils	1000–5000
Sand and gravel	50–1000
Surface limestone	100–10,000
Limestone	5–4000
Shales	5–100
Sandstone	20–2000
Granites, basalts, etc.	1000
Decomposed gneisses	50–500
Slates, etc.	10–100

for earthing design applications and vary only slightly in how the probes are spaced when compared to the Wenner method.

Electrical resistivity is the measurement of the specific resistance of a given material. It is expressed in ohm-meters and represents the resistance measured between two plates, covering opposite sides of a 1-m cube. This test is commonly

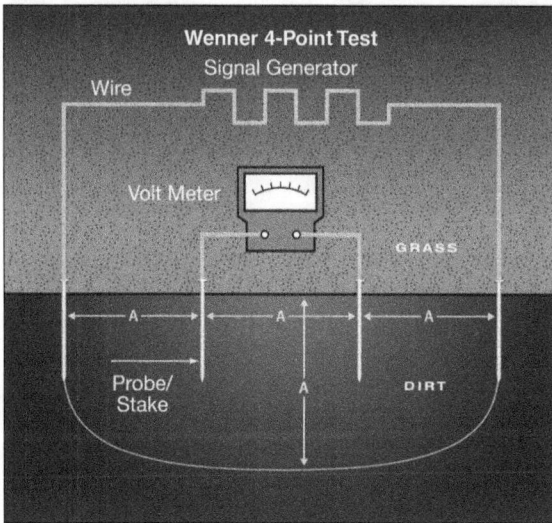

Figure 9.17 Four-point testing pattern.

performed at raw land sites, during the design and planning of earthing systems specific to the tested site. The test spaces four probes out at equal distances to approximate the depth of the soil to be tested. Typical spacings will be in the range of 0.3, 0.46, 0.6, 0.9, 1.2, 2.1, 9.84 m, and so on, with each spacing increasing from the preceding one, by a factor not greater than 1.5, up to a maximum spacing that is approximately 1–3 times the maximum diagonal dimension of the earthing system being designed. This results in a maximum distance between the outer current electrodes of 3–9 times the maximum diagonal dimension of the future earthing system. This is one "traverse" or set of measurements, and is typically repeated, albeit with shorter maximum spacings, several times around the location at right angles and diagonally to each other to ensure accurate readings.

The basic premise of the test is that probes spaced at 1.5 m distances across the surface, will measure the average soil resistivity to an approximate depth 1.5 m. The same is true if you space the probes 12 m across the earth, you get a weighted average soil resistance from 0 m down to 12 m in depth, and all points in between. This raw data must be processed with computer software to determine the actual resistivity of the soil as a function of depth.

Conducting a Wenner Four-Point Test. The following describes how to take one "traverse" or set of measurements. As the "four-point" indicates, the test consists of four pins that must be inserted into the earth. The outer two pins are called the current probes, C_1 and C_2. These are the probes that inject current into the earth. The inner two probes are the potential probes, P_1 and P_2. These are the probes that take the actual soil resistance measurement.

In the test shown in Fig. 9.18, a probe C_1 is driven into the earth at the corner of the area to be measured. Probes P_1, P_2, and C_2 are driven at 1.5, 3, and 4.6 m, respectively, from rod C_1 in a straight line to measure the soil resistivity from 0 to 1.5 m

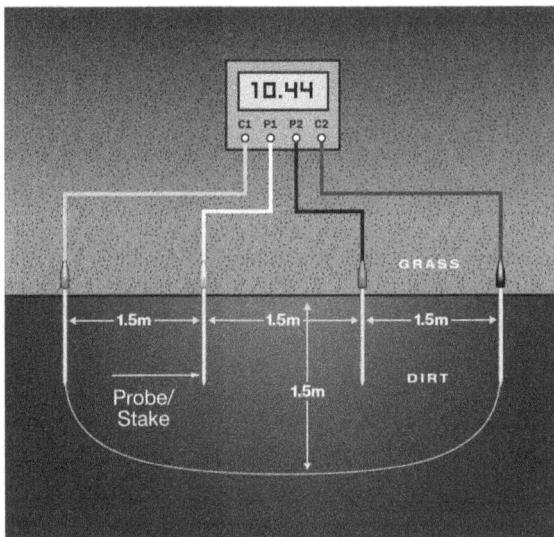

Figure 9.18 Four-point test setup.

in depth. C_1 and C_2 are the outer probes and P_1 and P_2 are the inner probes. At this point, a known current is applied across probes C_1 and C_2, while the resulting voltage is measured across P_1 and P_2. Ohm's law can then be applied to calculate the measured apparent resistance.

Probes C_2, P_1, and P_2 can then be moved out to 3, 20, and 30 m spacing to measure the resistance of the earth from 0 to 3 m in depth. Continue moving the three probes (C_2, P_1, and P_2) away from C_1 at equal intervals to approximate the depth of the soil to be measured. Note that the performance of the electrode can be influenced by soil resistivities, at depths that are considerably deeper than the depth of the electrode, particularly for extensive horizontal electrodes, such as water pipes, building foundations, or earthing grids.

Soil Resistance Meters. There are basically two types of soil resistance meters: Direct current (DC) and alternating current (AC) models, sometimes referred to as high-frequency meters. Both meter types can be used for four-point and three-point testing, and can even be used as a standard (two-point) voltmeter for measuring common resistances.

Care should always be given when selecting a meter, as the electronics involved in signal filtering are highly specialized. Electrically speaking, the earth can be a noisy place. Overhead power lines, electric substations, railroad tracks, various signal transmitters, and many other sources contribute to signal noise found in any given location. Harmonics, 50/60 Hz background noise, and magnetic field coupling can distort the measurement signal, resulting in apparent soil resistivity readings that are larger by an order of magnitude, particularly with large spacings. Selecting equipment with electronic packages capable of discriminating between these signals is critical.

Alternating current or high-frequency meters typically use a pulse signal operating at 128 pulses per second or greater. Often these AC meters claim to be using "pulsed DC" which is in reality simply a square-wave AC signal. These AC meters typically suffer from the inability to generate sufficient current and voltage (typically less than 50 mA and under 10 V) to handle long traverses and generally should not be used for probe spacings greater than 30.5 m. Furthermore, the high-frequency square-wave signal flowing in the current lead induces a noise voltage in the potential leads that cannot be completely filtered out: this noise becomes greater than the measured signal, as the soil resistivity decreases and the pin spacing increases. High-frequency meters are less expensive than their DC counterparts and are by far the most common meter used in soil resistivity testing.

Direct Current meters, which actually generate low-frequency pulses (on the order of 0.5–4.0 s/pulse), are the preferred equipment for soil resistivity testing, as they do away with the induction problem from which the high-frequency meters suffer. However, they can be very expensive to purchase. Depending upon the equipment's maximum voltage (500–2000 mA and 800 V peak to peak), DC meters can take readings with extremely large probe spacings and often many thousands of feet in distance. Typically, the electronics filtering packages offered in DC meters are superior to those found in AC meters. Care should be taken to select a reputable manufacturer.

Data Analysis. Once all the resistance data is collected, the following formula can be applied, to calculate the apparent soil resistivity in ohm-meter.

4-POINT DATA INTERPRETATION

$$\rho = 1.915\,A\,R$$
$$\rho = 1.915\,(40)\,(4.5)$$

$$\rho_a = \cfrac{4\pi AR}{1+\cfrac{2A}{\sqrt{(A^2+4B^2)}} - \cfrac{2A}{\sqrt{(4A^2+4B^2)}}}$$

ρ = Resistivity B = Depth of Probes A = Spacing of Probes R = Resistance (reading from meter)

If $A>20B$, then $\rho = 2\,\pi AR$ **= 1.915 AR**

Figure 9.19 Soil resistivity calculation.

For example, if an apparent resistance of 4.5 Ω is measured at 40-ft spacing, the soil resistivity in ohm-meter would be 344.7. Figure 9.19 shows the entire soil resistivity formula in detail. One refers to "apparent" resistivity, because this does not correspond to the actual resistivity of the soil. This raw data must be interpreted by suitable methods in order to determine the actual resistivity of the soil. Also note that the final 1.915 number is calculated by converting meters into feet.

Often when we describe a soil resistivity test, such as the Wenner four-point method, we correlate the spacings between the probes as a depth or sounding reading. In other words, the distance between the pins in theory equates to the approximate depth being measured. Remember that there are many factors that relate to the actual depth of the measurements read by the meter, so this concept is just a general guideline.

When a resistivity meter takes a measurement, that individual number means nothing in itself, without first doing a little math to determine resistivity. The simplified formula is to take the reading from the meter, multiply by the probe spacing (in feet), and then multiply it again by 1.915. The result is a resistivity number. When we combine a series of measurements taken at different spacings, we can begin to determine what the characteristics (resistivity) of the earth are like at various depths. The process of comparing numerous individual soil resistivity measurements (at differing spacings) is how one develops a soil model.

The soil model will show changes in resistivity of the earth at various depths. What is the resistivity of the soil at 1 m? What is it at 10 m? A good soil model will answer these questions. Of course, there are many rules as to how many measurements must be taken and at what spacings are required in order to get an accurate model, but that is a different topic. The concern in this case is what happens when a drastic change in resistivity occurs from one layer to the next.

When we conduct a soil resistivity test, we are injecting a test signal (electrical energy) into the surface of the earth, down through the soil to various depths, and recording the loss of energy as a resistance. As the electrical test signal passes from one layer to the next, the test signal will degrade in proportion to the changes in resistance that it encounters. This is especially true when the signal must try to move from a very conductive layer of soil to a very resistive layer of soil. The test signal will simply prefer to stay in the most conductive material.

If you have ever seen a submarine war movie, the sub commander will move his submarine below a colder layer of ocean water, to avoid being detected by sonar. The cold layer of water will bounce the sonar signal up and away from the submarine, hiding it from the enemy. This is similar to what happens when conducting soil resistivity tests; the test signal may in fact not penetrate the layers as well as we might hope.

These changes in layer resistivity affect the signal in a predictable way, and as such can be calculated and its effects corrected. This is why good engineers prefer soil resistivity models calculated using computer modelling programs instead of simple hand-calculations (good computer modelling programs perform thousands, if not hundreds of thousands, of calculations). Today's sophisticated algorithms take into account most of the variables and provide vastly superior and more accurate soil resistivity models.

That said, computer algorithms can only help correct the math. A good soil resistivity technician will know how to improve the original signal. The first step is to always use true DC meters. It is recommended to use 800-V p-p DC meters which require an additional car-battery to generate the needed power. The next step is to have many readings starting at 0.15-m spacings, with spacing intervals increasing at a factor of no greater than 1.5, with 1.33 preferred. You would also need to keep taking readings and increasing your spacings at those intervals until your spacings are at least as great as the depth you are trying to read, preferably two or three times as great. This would certainly mean many dozens of measurements and a lot of work.

Shallow Depth Readings. Shallow depth readings, as little as 6 in. depth, are exceedingly important for most, if not all, earthing designs. As described above, the deeper soil resistivity readings are actually weighted averages of the soil resistivity from the earth surface down to depth and include all the shallow resistance readings above it. The trick in developing the final soil model is to pull out the actual resistance of the soil at depth, and that requires "subtracting" the top layers from the deep readings. Figure 9.20 demonstrates how the shallowest readings impact deeper ones below it.

As you can see from Fig. 9.20, if you have a 1.5-m reading of 50 $\Omega \cdot$ m and a 3-m reading of 75 $\Omega \cdot$ m, the actual soil resistance from 1.5 to 3 m might be 100 $\Omega \cdot$ m (the point here is to illustrate a concept: precomputed curves or computer software are needed to properly interpret the data). The same follows true for larger pin spacings. The shallowest readings are used over and over again in determining the actual resistivity at depth.

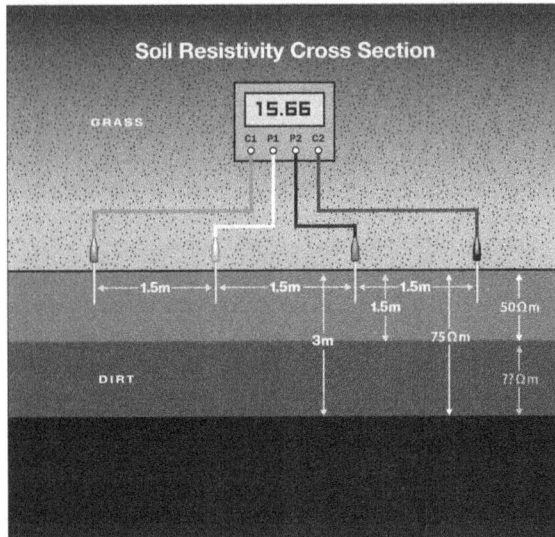

Figure 9.20 Importance of shallow readings.

Shallow depth readings of 0.15, 0.3, 0.46, 0.6, and 0.76 m are important for earthing design, because earthing conductors are typically buried at 0.46–0.76 m below the surface of the earth. To accurately calculate how those conductors will perform at these depths, shallow soil readings must be taken. These shallow readings become even more important when engineers calculate EPR, touch voltages, and step voltages.

It is critical that the measurement probes and current probes be inserted into the earth to the proper depth for shallow soil resistivity readings. If the probes are driven too deep, then it can be difficult to resolve the resistivity of the shallow soil. Ideally a 20 to 1 ratio (5%) is best; however, when doing very shallow readings that rule cannot always apply. A good rule of thumb when conducting short spacing measurements is that the penetration depth of the potential probes should be no more than 10% of the pin spacing, whereas the current probes must not be driven more than 30% of the pin spacing.

Deep Readings. Often, the type of meter used determines the maximum depth or spacing that can be read. A general guideline is that alternating current soil resistivity meters are good for no more than 30.5-m pin spacings, particularly in low-resistivity soils. For greater pin spacings, DC soil resistivity meters are required. While direct current meters are always the preferred choice, they are the only type of meter that can generate the required voltages needed to push the signal through the soil at long distances free of induced voltages and signal noise error from the current injection leads.

Test Location. Soil resistivity testing should be conducted as close to the proposed earthing system as possible, taking into consideration the physical items that may cause erroneous readings. There are two issues that may cause poor quality readings:

1. Electrical interference, causing unwanted signal noise to enter the meter.
2. Metallic objects "short-cutting" the electrical path from probe to probe. The rule of thumb here is that a clearance equal to the pin spacing should be maintained, between the measurement traverse and any parallel-buried metallic structures.

Testing in the vicinity of the site in question is obviously important; however, it is not always practical. Many electric utility companies have rules regarding how close the soil resistivity test must be in order to be valid. The geology of the area also plays into the equation, as dramatically different soil conditions may exist only a short distance away.

When left with little room or poor conditions in which to conduct a proper soil resistivity test, one should use the closest available open field, with as similar geological soil conditions as possible.

Testing of Existing Earthing Systems.   The measurement of earth resistance for an existing earth electrode system is very important. It should be done when the electrode is first installed and then at periodic intervals thereafter. This ensures that the RTE does not increase over time. There are two methods for testing an existing earth electrode system. The first is the three-point or fall-of-potential method and the second is the induced frequency test or clamp-on method. The three-point test requires complete isolation from the power utility (see Fig. 9.21). Not just power isolation, but also removal of any neutral or other earth connections extending outside the earthing system. This test is the most suitable test for large earthing systems and single isolated earthing electrodes, when initially constructed.

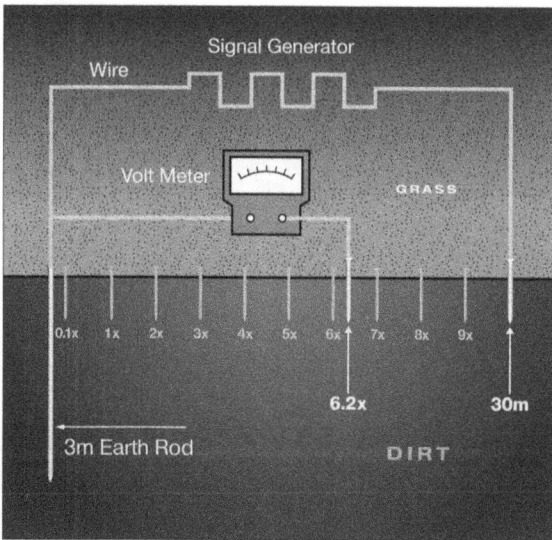

Figure 9.21 Three-point test method.

The second method is the induced frequency test and can be performed while power is on. It actually requires the utility to be connected to the earthing system under test. This test is accurate only for small electrodes, as it uses frequencies in the kilohertz range, which see long conductors as inductive chokes and therefore do not reflect the 50/60 Hz resistance of the entire earthing system.

Both tests inject a signal into the electrode system; they differ only in the return. The three-point test uses a small probe, installed at some distance from the electrode, as the signal return; the-induced frequency test uses the utility company's earthing system.

Fall-of-Potential Method or The Three-Point Test. The three-point or fall-of-potential method is used to measure the RTE of existing earthing systems. The two primary requirements, to successfully complete this test are the ability to isolate the earthing system from the utility neutral and knowledge of the diagonal length of the earthing system, (i.e., a 1.8 m × 2.4 m earthing ring would have a 3 m diagonal length). In this test, a short probe, referred to as probe Z, is driven into the earth, at a distance of 10 times the diagonal length of the earthing system (rod X). A second probe (Y) is placed in-line, at a distance from rod X equal to the diagonal length of the earthing system (see Fig. 9.22).

At this point, a known current is applied across X and Z, while the resulting voltage is measured across X and Y. Ohm's Law can then be applied ($R = V/I$) to calculate the measured resistance. Probe Y is then moved out to a distance of two times the diagonal length of the earthing system, in-line with X and Z, to repeat the resistance measurement at the new interval. This will continue, moving probe Y out to three times, four times ... nine times, the diagonal length, to complete the three-point test, with a total of nine resistance measurements.

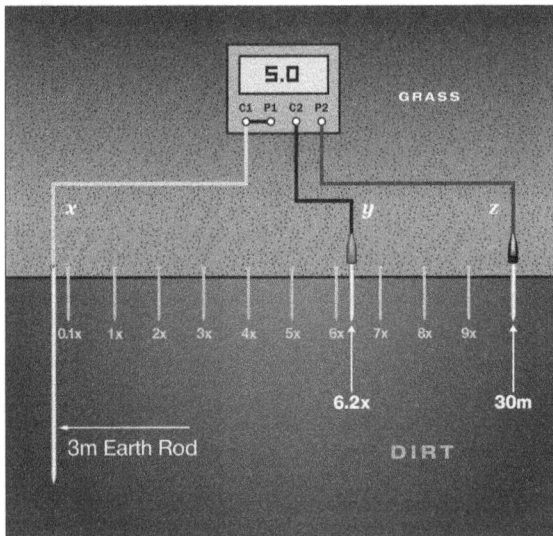

Figure 9.22 Three-point test setup.

Graphing and Evaluation. The three-point test is evaluated, by plotting the results as data points with the distance from rod X along the X-axis and the resistance measurements along the Y-axis, to develop a curve. Roughly midway between the centre of the electrode under test and the probe Z, a plateau or "flat spot" should be found, as shown in the graph. The resistance of this plateau, (actually, the resistance measured at the location 62% from the centre of the electrode under test, if the soil is perfectly homogeneous) is the RTE of the tested earthing system.

Invalid Tests. If no semblance of a plateau is found and the graph is observed to rise steadily, the test is considered invalid. This can be due to the fact that probe Z was not placed far enough away from rod X, and can usually indicate, that the diagonal length of the earthing system was not determined correctly. If the graph is observed to have a low plateau, that extends the entire length and only rises at the last test point, this may also be considered invalid (see Fig. 9.23). This is often because the utility or telecom neutral connection remains connected to the earthing system.

Induced Frequency Testing or Clamp-On Testing. The induced frequency testing or commonly called *clamp-on* testing is one of the newest test methods for measuring the RTE of an earthing system or electrode. This test uses a special transformer to induce an oscillating voltage (often 1.7 kHz) into the earthing system. Unlike the three-point test, which requires the earthing system to be completely disconnected and isolated before testing, this method requires that the earthing system under test be connected to the electric utility earth system (or other large earthing systems such as one from the telephone companies), with a connecting conductor (typically via the neutral return wire) to provide the return path for the signal. This test is the only test that can be used on live or "hot" systems. However, there are some limitations, primarily that being:

1. The amount of amperage running through the tested system must be below the equipment manufacturer's limits.

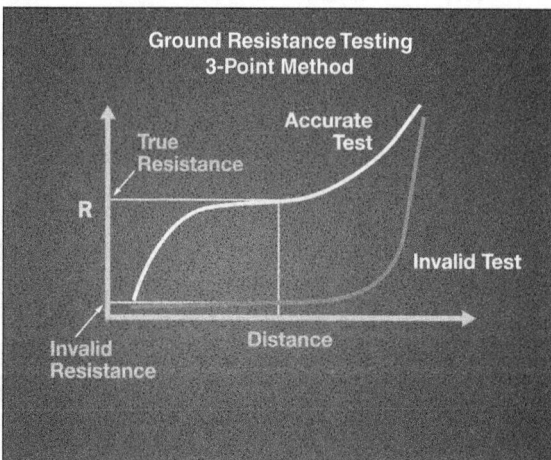

Figure 9.23 Three point fall of potential graph.

2. The test signal must be injected at the proper location so that the signal is forced through the earthing system and into the earth.
3. The instrument a *clamp-on* meter actually measures the sum of the resistance of the earthing system under test, and the impedance of the utility neutral earthing, including the neutral wiring. Due to the high frequency used, the impedance of the neutral wiring is non-negligible and can be greater than the earth resistance of a very low-resistance earthing system under test, which can therefore not be measured accurately.
4. The earth resistance of a large earthing system at 50/60 Hz can be significantly lower than at 1.7 kHz.

Many erroneous tests have been conducted where the technician only measured metallic loops and not the true RTE of the earthing system. The veracity of the induced frequency test has been questioned due to testing errors. However, when properly applied to a small-to-medium-sized, self-standing earthing system, this test is rapid and reasonably accurate.

Test Application.   The proper use of this test method requires the utility neutral to be connected to an earthed wye-type transformer. The oscillating voltage is induced into the earthing system at a point where it will be forced into the soil and return through the utility neutral. Extreme caution must be taken at this point, as erroneous readings and mistakes are often made. The most common of these occur when clamping on or inducing the oscillating voltage into the earthing system at a point where a continuous metallic path exists back to the point of the test. This can result in a continuity test being performed rather than an earth resistance test. Understanding the proper field application of this test is vital to obtaining accurate results. The induced frequency test can test earthing systems that are in use and does not require the interruption of service to take measurements (see Fig. 9.24).

Earth Resistance Monitoring.   Earth resistance monitoring is the process of automated, timed, and/or continuous RTE measurement. These dedicated systems use

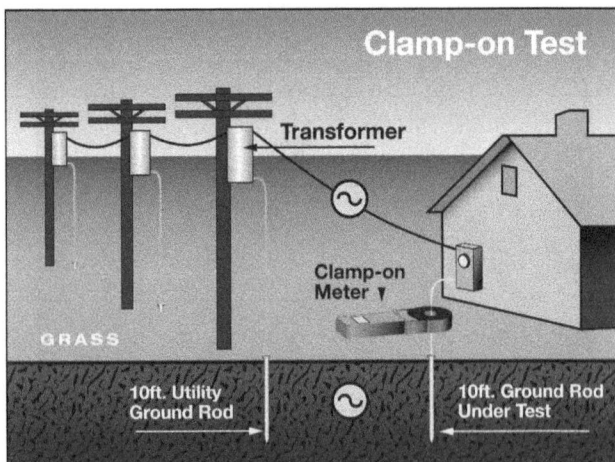

Figure 9.24 Induced frequency test diagram.

the induced frequency test method to continuously monitor the performance of critical earthing systems. Some models may also provide automated data reporting. These new meters can measure RTE and the current that flows on the earthing systems that are in use. Another benefit is that it does not require interruption of the electrical service to take these measurements.

EARTH POTENTIAL RISE

Earth potential rise (EPR) is a phenomenon that occurs when large amounts of electricity enter the earth. This typically happens when substations or high-voltage towers fault, or when lightning strikes occur. When currents of large magnitude enter the earth from an earthing system, not only will the earthing system rise in electrical potential, but so will the surrounding soil as well.

The voltages produced by an EPR event can be hazardous to both personnel and equipment. As described earlier, soil has resistance which will allow an electrical potential gradient or voltage drop to occur along the path of the fault current in the soil. The resulting potential differences will cause currents to flow into any and all nearby earthed conductive bodies, including concrete, pipes, copper wires, and people.

Earth Potential Rise (EPR) Definitions

Earth potential rise (as generally defined in IEEE Std. 367) is the product of an earth electrode impedance, referenced to remote earth, and the current that flows through that electrode impedance.

Earth Potential Rise (as generally defined by IEEE Std. 80-2000) is the maximum electrical potential that a (substation) earthing grid may attain, relative to a distant earthing point, assumed to be at potential of remote earth. This voltage, EPR, is equal to the maximum grid current times the grid resistance.

EPR events are a concern wherever electrical currents of large magnitude flow into the earth. This can be at a substation, high-voltage tower or pole, or a large transformer. In cases where an EPR event may be likely, earthing precautions are required to ensure personnel and equipment safety.

Electrical potentials in the earth drop abruptly around the perimeter of an earthing system, but do not drop to zero. In fact, in a perfectly homogeneous soil, soil potentials are inversely proportional to the distance from the centre of the earthing system, once one has reached a distance that is a small number of earthing system dimensions away. The formula is as follows:

Maximum earth potential = soil resistivity × current/(2π × distance)

Where earth potential is in volts, soil resistivity is in ohm-meters, current is the current flowing into the soil from the earthing system in amperes, circle constant pi (unit less), and distance is in meters from source current.

Probably the most commonly noted EPR event involves the death of cows in a field during a lightning strike. Imagine lightning striking the centre of an open field where cows are standing. The current injected into the earth flows radially away from the strike point, in all directions, creating voltage gradients on the surface of the earth, also in a radial direction. All the cows facing the lightning strike would have their fore hooves closer to the strike point than their rear hooves. This would result in a difference of potential between their fore and rear legs, causing current to flow through their bodies, including the heart, and killing the cow. On the other hand, those cows with their flanks turned toward the lightning strike would have a greater chance of surviving, as the distance between their forelegs and therefore the voltage applied between them, would be relatively small, resulting in a lesser current flow.

EPR studies are typically conducted on substations and high-voltage towers. Substations have relatively large earthing areas, especially when compared to high-voltage towers and poles. Towers and poles represent by far the most potentially dangerous and difficult EPR situations to handle and are often not protected, unless they are located in high-exposure areas or have equipment installed at earth level at which service personnel might be required to work.

Earth Potential Rise Analysis. The primary purpose of an EPR Study is to determine the level of hazard associated with a given high-voltage location, for personnel and/or equipment. When the degree of hazard is identified, the appropriate precautions must be made to make the site safe. To do this, the engineer must identify what the minimum earthing system for each location will be. The engineer must also take into consideration all local and federal guidelines, including utility company requirements.

For example: Many utility companies require at a minimum that a simple earth ring be installed at least 0.46 m below earth and 1 m from the perimeter of all metal objects. This earth ring is also referred to as a counterpoise.

Once the minimum earthing system is identified, the engineer can run an EPR analysis and identify the extent of any electrical hazards (see Fig. 9.25).

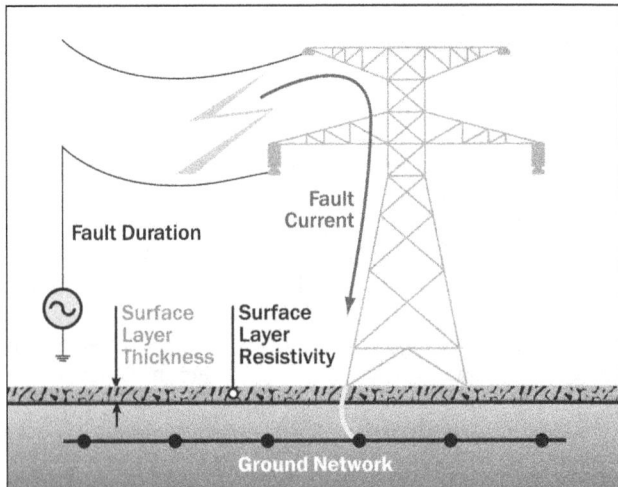

Figure 9.25 Electric fault at a transmission tower.

Typically, items reported in a EPR Study will include the following: the square footage, size and layout of the proposed earthing grid, RTE of the proposed earthing system, the estimated fault current that would flow into the earthing system, EPR (in volts) at the site, 300-V Peak line, the X/R Ratio, and the fault clearing time in seconds. Touch and step voltages are usually computed as well, as these are the primary indicators of safety.

The earthing engineer needs three pieces of information to properly conduct an EPR Study:

1. Soil resistivity data
2. Site drawings with the proposed construction
3. Electrical data from the power company

Soil Resistivity Data. The soil resistivity data should include apparent resistivity readings at pin spacings ranging from 0.15 m to as many as three earthing grid diagonals, if practical. Touch and step voltages represent the primary concern for personnel safety. Understanding the characteristics of the soil, at depths ranging from immediately underfoot to one or more grid dimensions, is required for a cost-effective and safe earthing system to be designed. See previous section regarding soil resistivity.

Site Drawings. The proposed site drawings should show the layout of the high-voltage tower or substation, and any additional construction for new equipment that may be occurring on the site, including fencing and gate radius. Incoming power and Telco runs should also be included. In the case of high-voltage towers, the height and spacing of the conductors carried on the tower, and any overhead earth wires that may be installed on the tower, need to be detailed during the survey. This information is needed to properly address all the touch and step voltage concerns that may occur on the site.

Electric Utility Data. The electric utility company needs to provide electrical data regarding the tower or substation under consideration. This data should include the name of the substation or the number of the tower, the voltage level, the sub-transient X/R ratio, and the clearing times. In the case of towers, the line names of the substations involved, the amount of current contributed by each substation in the event of a fault, and the type and positions of the overhead earth wires, if any, with respect to the phase conductors installed on each tower or pole. If overhead earthing wires are present, tower or pole earth resistances along the line are of interest as well, be they measured, average, or design values.

This information is important, as high-voltage towers have small earth area, yet handle very large amounts of electricity. Knowing if a tower has an overhead earth wire is important, because the overhead wire will carry away a percentage of the current, which will depend on the overhead earth wire type and earth resistances of adjacent towers, to other towers in the run, reducing the EPR event. Additionally, towers with overhead earth wires tend to have shorter clearing times. The same holds for substations: overhead earth wires on transmission lines and neutral wires on distribution lines can significantly reduce the magnitude of fault current, which flows into the substation earthing system during fault conditions.

The following information is required from the utility company:

1. Phase-to-earth fault current contributed by each power line circuit
2. Fault clearing time
3. Line voltage

4. Sub-transient X/R ratio
5. The make/type/number of overhead earth wires on each tower/pole line and position with respect to the phase conductors
6. Earth wire continuity and bonding configuration back to the tower and substation
7. The average distance from tower-to-tower and tower-to-substation
8. Typical tower/pole earth resistance: measured or design values

As-built drawings are often acquired and are useful for towers with existing earthing systems. They are also useful in the case of modifications and upgrades to existing substations, which will have extensive earthing systems already installed.

Personnel Safety during Earth Potential Rise Events. The earthing engineer will be required to develop safety systems to protect any personnel working where EPR hazards are known to exist. Many laws and regulations mandate that all known hazards must be eliminated from the workplace for the safety of workers. BS 7354, IEEE Std. 80, and ENA TS 41-2 specify the requirements around how step and touch potentials are to be eliminated on hot sites, which may include any related communication equipment associated with a hot site.

Substations are always considered workplaces and step and touch potentials (voltages) must always be eliminated. Transmission and distribution towers or poles are not always considered workplaces and therefore are often exempt from these requirements. Take, for example, a lonely tower on a mountain side, or in the middle of the desert: these towers are not typically considered workplaces. However, any high-voltage tower or pole becomes a workplace, as soon as equipment is installed that is not related to the electric utility company and requires outside vendors to support the new equipment. Cellular telecommunications, environmental monitoring, and microwave relay equipment are good examples of equipment that, when installed on a high-voltage tower, turns the tower into a workplace. This would make the elimination of step and touch potentials required.

Hazardous Voltages. Fibrillation current is the amount of electricity needed to cause cardiac arrest from which recovery will not spontaneously occur in a person and is a value based on statistics. *BS 7354, BS IEC 60050-195, IEV 195.05.11 modified and − C12, IEEE Std. 80, ENA TS 41-2, and IEC Report 479-1* provide a method to determine the pertinent value of fibrillation current for a safety study, along with a good explanation of how it is derived. Many different methods exist for calculating fibrillation current; however the 50-kg IEEE method is the most commonly used in North America. The formula used shows that the fibrillation current level is inversely proportional to the square root of the fault duration; however, it must be increased by a correction factor, based on the sub-transient X/R ratio, which can be quite large for shorter fault durations. If personnel working at a site during fault conditions experience voltages that will cause a current less than the fibrillation current to flow in their bodies, then they are considered safe. If a worker will experience a greater voltage than is acceptable, additional safety precautions must be taken.

The sub-transient X/R ratio at the site of the fault is important in calculating the acceptable fibrillation current and to determine the maximum allowable step and touch potentials that can occur at any given site.

Fault duration is a required piece of data for properly calculating maximum allowable step and touch potentials. The fault duration is the amount of time it takes for the power company to shut off the current in the event of a fault.

Ultimately the engineer must determine two things:
1. The site-specific maximum allowable voltage that a person can safely withstand
2. The actual voltages that will be experienced at the site during a fault

Each site will have different levels of voltages for both of the above. Unfortunately, we cannot simply say that a human being can withstand a certain level of current or voltages and use that value all the time, since the maximum safe human voltage threshold is determined by the surface layer resistivity, the fault duration, and the sub-transient X/R ratio. Additionally as each site has different fault durations and different soil conditions, it is critical that calculations be made for each and every possible fault location.

Step Potential. When a fault occurs at a tower or substation, the current will enter the earth. Based on the distribution of varying resistivity in the soil (typically, a horizontally layered soil is assumed) a corresponding voltage distribution will occur. The voltage drop in the soil surrounding the earthing system, can present hazards for personnel standing in the vicinity of the earthing system. Personnel "stepping" in the direction of the voltage gradient could be subjected to hazardous voltages.

In the case of step potentials, electricity will flow if a difference in potential exists between the two legs of a person (see Fig. 9.26). Calculations must be performed that determine how great the tolerable step potentials are and then compare those results to the step voltages expected to occur at the site.

Hazardous step potentials can occur a significant distance away from any given site. The more current that is pumped into the earth, the greater the hazard. Soil

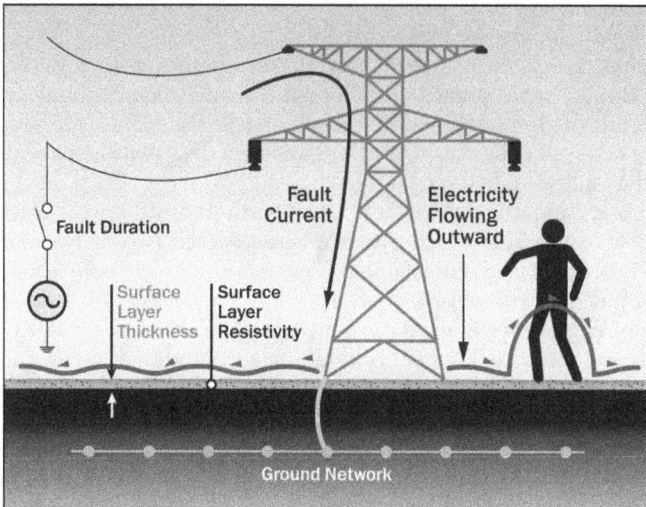

Figure 9.26 Step potential near a transmission tower.

resistivity and layering plays a major role in how hazardous a fault occurring on a specific site may be. Low-over-high soil models tend to increase step potentials at further distances from the earth fault location. While high-over-low soil models tend to increase step potentials when in close proximity to the earth fault location.

A high-resistivity top layer and low-resistivity bottom layer tends to result in the highest step voltages when personnel are in close proximity to the source of the earth fault: the low-resistivity bottom layer draws more current out of the electrode through the high-resistivity layer, resulting in large voltage drops near the electrode.

When personnel are standing further from the source of the earth fault, the worst-case scenario occurs when the soil has conductive top layers and resistive bottom layers: in this case, the fault current remains in the conductive top layer for much greater distances away from the electrode.

Fault clearing time is an important factor to consider as well. The more time it takes the electric utility company to clear the fault, the more likely it is for a given level of current to cause the human heart to fibrillate.

It is important to remember that step voltages are life-threatening because most power companies use automated re-closers. In the event of a fault, the power is shut off and then automatically turned back on. This is done in case the faults occurred due to an unfortunate bird that made a poor choice in where to rest, or dust that may have been burned off during the original fault. A few engineers believe that fibrillation current for step potentials must be far greater than touch potentials, as current will not pass through any vital organs in the former case. This is not always true, as personnel that receive a shock due to step potentials may fall to the earth, only to be hit again, before they can get up, when the automatic re-closers activate.

Touch Potentials.   When a fault occurs at a tower or substation, the current will pass through any metallic object and enter the earth. Those personnel "touching" an object in the vicinity of the EPR will be subjected to these voltages which may be hazardous.

For example, if a person happens to be touching a high-voltage tower leg when a fault occurs, the electricity would travel down the tower leg into the person's hand and through vital organs of the body. It would then continue on its path and exit out through the feet and into the earth (see Fig. 9.27). Careful analysis is required to determine the acceptable fibrillation currents that can be withstood by the body if a fault were to occur.

Engineering standards use a 1-m (3.23-ft) reach distance for calculating touch potentials. A 2-m (6.54-ft) reach distance is used when two or more objects are inside the EPR event area. For example, a person could have both arms stretched out and be touching two objects at once such as a tower leg and a metal cabinet. Occasionally, engineers will use a three-meter distance to be particularly cautious, as they assume someone may be using a power tool with a power cord 3 m in length.

The selection of where to place the reference points used in the touch potential calculations are critical in getting an accurate understanding of the level of hazard at a given site. The actual calculation of touch potentials uses a specified object (such as a tower leg) as the first reference point. This means that the further away from the tower the other reference point is located, the greater the difference in

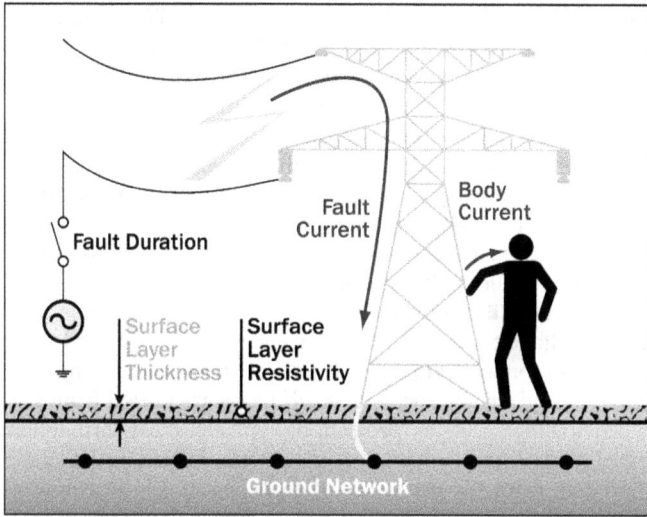

Figure 9.27 Touch potential at a transmission tower.

potential. If you can imagine a person with incredibly long arms touching the tower leg and yet standing many dozens of feet away, you would have a huge difference in potential between their feet and the tower. Obviously, this example is not possible: this is why setting where and how far away the reference points used in the touch calculation, is so important and why the 1-m rule has been established.

Mitigating Step and Touch Potential Hazards. Mitigating step and touch potential hazards is usually accomplished through one or more of the following 3 main techniques:
1. Reduction in the RTE of the earthing system
2. Proper placement of earth conductors
3. The addition of resistive surface layers
Understanding the proper application of these techniques is the key to reducing and eliminating any EPR hazards. Only through the use of highly sophisticated 3-dimensional electrical simulation software that can model soil structures with multiple layers and finite volumes of different materials, can the engineer accurately model and design an earthing system, which will safely handle high-voltage electrical faults.

Reduce the Resistance to Earth. Reducing the RTE of the site is often the best way to reduce the negative effects of any EPR event, where practical. The EPR is the product of the fault current, flowing into the earthing system, times the RTE of the earthing system; in essence Ohm's Law. Thus, reducing the RTE will reduce the EPR to the degree that the fault current flowing into the earthing system does increase in response to the reduced RTE. For example, if the fault current for a high-voltage tower is 5000 A and the RTE of the earthing system is 10 Ω, the EPR will be 50,000 V. If we reduce the RTE of the earthing system down to 5 Ω and the fault current increases to 7000 A as a result, then the EPR will become 35,000 V.

As seen in the example above, the reduction of the RTE can have the effect of allowing more current to flow into the earth at the site of the fault, but will always result in lower EPR values and touch and step voltages at the fault location. On the other hand, further away from the fault location, at adjacent facilities not connected to the faulted structure, the increase in current into the earth will result in greater current flow near these adjacent facilities and therefore an increase in the EPR, touch voltages, and step voltages at these facilities. Of course, if these are low to begin with, an increase may not represent a problem, but there are cases in which a concern may exist. Reducing the RTE can be achieved by any number of means, as discussed earlier in this chapter.

Proper Placement of Earth Conductors. A typical specification for earth conductors at high-voltage towers or substations is to install an earth loop around all metallic objects and connected to the objects; keep in mind that it may be necessary to vary the depth and/or distance that earth loops are buried from the structure in order to provide the necessary protection. Typically these earth loops require a minimum size of 10.60 mm bare copper conductor, buried in direct contact with the earth and 1 m from the perimeter of each object, 0.46 m below grade. The purpose of the loop is to minimize the voltage between the object and the earth surface, where a person might be standing while touching the object: that is, to minimize touch potentials.

It is important that all metallic objects in an EPR environment be bonded to the earth system to eliminate any difference in potentials. It is also important that the resistivity of the soil, as a function of depth, be considered in computed touch and step voltages and in determining at what depth to place conductors. For example, in a soil with a dry, high-resistivity surface layer, conductors in this layer will be ineffective; a low-resistivity layer beneath that one would be the best location for earthing conductors. On the other hand, if another high-resistivity layer exists further down, long earth rods or deep wells extending into this layer will be less effective.

It is sometimes believed, that placing horizontal earthing loop conductors very close to the surface, results in the greatest reduction in touch potentials. This is not necessarily so, as conductors close to the surface are likely to be in drier soil, with a higher resistivity, thus reducing the effectiveness of these conductors. Furthermore, while touch potentials immediately over the loop may be reduced, touch potentials a short distance away may actually increase, due to the decreased zone of influence of these conductors. Finally, step potentials are likely to increase at these locations: indeed, step potentials can be a concern near conductors that are close to the surface, particularly at the perimeter of an earthing system. It is common to see perimeter conductors around small earthing systems, buried to a depth of 1 m below grade, in order to address this problem.

Resistive Surface Layers. One of the simplest methods of reducing step and touch potential hazards is to wear electric hazard shoes. When dry, properly rated electric hazard shoes have millions of Ω of resistance in the soles and are an excellent tool for personnel safety. On the other hand, when these boots are wet and dirty, current may bypass the soles of the boots, in the film of material that has accumulated on the sides of the boot. A wet leather boot can have a resistance as low as 100 Ω. Furthermore, it cannot be assumed that the general public, who may have access to the outside perimeter of some sites, will wear such protective gear.

Another technique used in mitigating step and touch potential hazards is the addition of more resistive surface layers. Often a layer of crushed rock is added to a tower or substation to provide a layer of insulation between personnel and the earth. This layer reduces the amount of current that can flow through a given person and into the earth. Weed control is another important factor, as plants become energized during a fault and can conduct hazardous voltages into a person. Asphalt is an excellent alternative, as it is far more resistive than crushed rock, and weed growth is not a problem. The addition of resistive surface layers always improves personnel safety during an EPR event. Please review the previous sections on step and touch potentials.

Telecommunications in High-Voltage Environments.   When telecommunications lines are needed at a high-voltage site, special precautions are required to protect switching stations from unwanted voltages. Running any copper wire into a substation or tower is going to expose the other end of the wire to hazardous voltages and certain precautions are required.

Industry standards regarding these precautions and protective requirements are covered in IEEE Standards 387, 487, and 1590. These standards require that an EPR study be conducted so that the 300-V peak line can be properly calculated (see Fig. 9.28).

To protect the telephone switching stations, telecommunication standards require that fiber-optic cables be used instead of copper wires within the 300-V peak line. A copper-to-fibre conversion box must be located outside the EPR event area at a distance in excess of the 300-V peak which is the same as the 212-V RMS line. This is known in the industry as the *300-V line*. This means that based on the calculation results, copper wire from the telecommunications company may

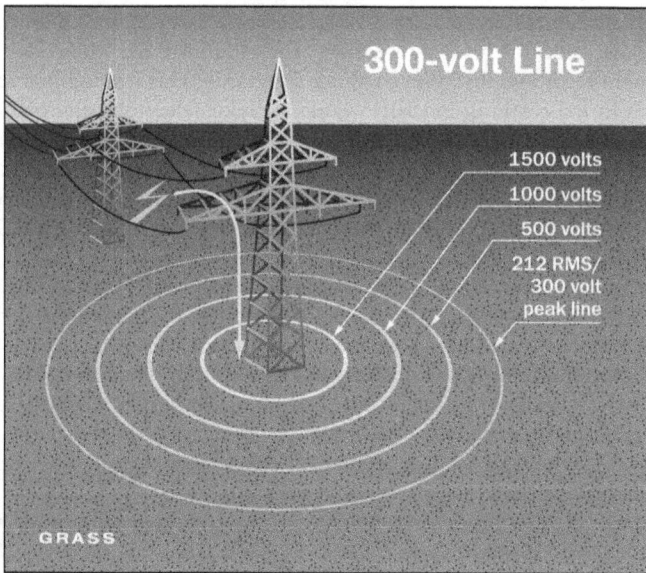

Figure 9.28 300-V line at a transmission tower.

not come any closer than the 300-V peak distance. This is the distance where copper wire must be converted over to fibre-optic cable. This can help prevent any unwanted voltages from entering the phone companies' telecommunications network.

The current formulae for calculating the 300-V line, as listed in the standards, has led to misinterpretation and divergences of opinion, resulting in order-of-magnitude, variations in calculated distances for virtually identical design input data. Furthermore, operating experience has shown that a rigorous application of theory results in unnecessarily large distances. This has caused many compromises within the telecommunications industry. The most noted one is a newer standard, *IEEE Std. 1590-2003*, that lists a 150-m (~500-ft) mark as a default distance, if an EPR study has not been conducted at a given location; although it is always recommended to calculate the actual voltage and not use this short cut.

Regulations and Standards for Human Safety in High-Voltage Environments. The primary regulation and enforcement for human safety in high-voltage environments is under *BS 7354, IEEE Std. 80, and ENA TS 41-2*. These regulations require the elimination of known electrical hazards in the workplace, including hazardous step and touch voltages. The law does not inform you how to eliminate these electrical hazards or what standards should be used; it merely requires you do eliminate them.

IEEE Std. 80 is the *IEEE Guide for Safety in AC Substation Earthing*. It was originally published in 1961 and has been updated three times, in 1976, 1986, and finally in 2000. This standard is fairly difficult to understand mostly not only because of the complex nature of the topic, but because the standard is trying to accomplish two functions. First, it is trying to provide a reasonably safe environment for personnel working in high-voltage substations. Second, it is trying to show how to calculate and mitigate hazardous step and touch voltages.

There are numerous areas outside of substations that require the mitigation of hazardous step and touch voltages, and as such the IEEE Std. 80 is often listed as a requirement for non-substation earthing standards. This has led to a lot of confusion with in the earthing industry. When cellular telephone companies install equipment on high-voltage transmission towers, the tower must be made safe for the cellular personnel. The only available standard for mitigating these hazardous voltages is the IEEE Std. 80, which specifically states it is only applicable to substations. As a result many companies ignore the requirements of IEEE Std. 80, placing their own personnel at risk.

An excellent solution for the IEEE would be if they split Standard 80 into two separate standards, one for the protection of personnel in substations and another for the mitigation of hazardous step and touch voltages.

Computer Modelling for Human Safety in High-Voltage Environments. Back in the early 1960s, when the first standards were being developed to mitigate the hazards to human safety caused by step and touch voltages, the best that any engineer could do was go through a series of hand-calculations and try and determine the best ways to make a particular high-voltage environment safe. Today, we have the ability to use computer modelling which can take into account far more variables than could ever be reasonably calculated by hand. The goal of this section is to make the argument that computer modelling is the only cost-effective and ethical way to comply with the requirements of BS 7354, IEEE Std. 80 and ENA TS 41-2.

There are a number of engineering programs on the market that can be useful in designing safe workplaces in high-voltage environments. These programs generally fall into one of two categories: those that provide a general analysis of the hazards, and those that provide an accurate detailed analysis. The differences between these two may sound small, but in reality it is quite important. The general analysis programs usually result in over installation in earthing in areas that don't need it, and the under installation in earthing in the areas that do need it. The reason for this is that the general analysis programs do not calculate step and touch voltages from specific objects within the computer model, but use an average potential across the entire compound.

The computer programs that provide a detailed analysis of the step and touch voltage hazards at a given site are typically quite expensive and require a great deal of training to master. The simulations conducted by these systems can take hours and hours of computing time using high-speed computers, just to run a single simulation. This is in stark contrast to the general analysis systems that provide near instantaneous results on even the slowest of computers. Detailed computer simulation programs are also capable of using soil models with three or more layers.

A proper study of the step and touch voltage hazards found at a given site involves a detailed drawing of all of the metallic objects at the site, including transformers, towers, switches, concrete foundations, buildings, posts, fences, gates/doors, and any other object that could be touched by a person, along with the earthing system. These objects must be placed in a multilayered soil model. Most sites require soil models with three to five layers in order to accurately model the propagation of electrical energy through the earthing system and into the earth. Single and two-layer models are almost always inaccurate. Single layer soils simply do not exist in reality, and two-layer models typically present such a high contrast between the layers as to present near impossible to overcome calculation errors. The general analysis computer programs are almost always limited to two-layer soil models and often have further limitations as to the available resistivities and depths that they can calculate.

Chapter 13 of the 2000 edition of IEEE Std. 80 deals with the soil structures and the selection of a soil model. The new edition introduced a flawed method for the calculation of a two-layer soil structure, and only briefly discusses multilayer structures. This new method uses a uniform soil structure that claims to provide accurate two-layer soil equivalences. In a paper titled *Effects of the Changes in IEEE Std. 80 on the Design and Analysis of Power System Earthing*, J. Ma, F. Dawalibi, and R. Southey, of Safe Engineering Services & Technologies, state that the two-layer soil structure method presented in Annex E of the IEEE 80 standard is flawed. The paper shows that in some cases when using the IEEE Std. 80 2-layer soil method, when the earthing grid is increased in size, the calculated RTE increased. This is contrary to what will happen in reality, when an earthing grid is increased in size, the RTE of the grid will decrease. In other examples, when the soil resistivity was increased in the two-layer calculation, the resistance of the earthing grid was calculated to decrease. Again, this is contrary to what will occur in real life. These results of the paper demonstrate that the calculations found in IEEE Std. 80 are only valid for uniform (single-layer) models and should not be used for multilayer soil configurations. This further demonstrates the need for detailed computer models, instead of the suggested hand calculations from annex E of Std. 80.

But even beyond the issue of being able to accurately calculate multilayer soil models, there are still far too many variables to ever calculate by hand. There are the formulas regarding permissible body current that must be analyzed, new requirements for calculating the surface layer Derating Factor, the calculation of foot resistances is now based on a formula with a rigorous series expression with each term being a surface integral, one must also take into account the effects of the DC offset current generated due to the asymmetrical fault current, to name only a few of the basics. Combining all of this data into trying to determine the touch voltage for a 1-m reach distance at a 45° angle from a specific transformer located at a specific location within an earth grid at a substation on top of a three-layer soil model, is nearly impossible to do by hand. And that is just one location around a single transformer that may have dozens of touchable locations around its perimeter! Let alone the thousands, if not hundreds of thousands of other locations that could be touched within the substation that must also be analyzed.

There are numerous other issues (particularly calculating the touch voltages for the exterior fences) that could be discussed, but by now it should be obvious that hand calculating these variables would be a daunting task. In fact, if you were to print out all the calculations that a detailed computer model performs, it would look like a doctorial thesis. In short, computer modelling is the only viable and ethical way to design a substation earthing grid. In today's computer age and under *BS 7354, IEEE Std. 80, and ENA TS 41-2*, a jury may consider hand calculations and/or general computer modelling to be criminal negligence. Detailed computer modelling is the only way to accurately conduct the proper *IEEE Std. 80* calculations.

THE EFFECTS OF LIGHTNING ON AN EARTHING SYSTEM

Lightning is an atmospheric electrostatic discharge (spark) which can travel at speeds of 220,000 km/h and can reach temperatures approaching 30,000°C which is hot enough to fuse silica sand into glass. There are some 50–100 lightning strikes occurring somewhere on the planet every second, can carry as much as 200,000 A of electricity and generate over 100 million V with each strike. It is estimated that there are over 300,000 lightning strikes in Britain each year, and that there are 30–60 severe injuries with an average of three deaths annually from lightning. In fact, in June of 2014, Britain was struck over 100,000 on a single weekend!

There are a number of engineering issues to be concerned with when lightning strikes an object, including a power plant. For this discussion, we will assume that this is a basic power plant/substation with an overhead lightning protection system (LPS) tied into the standard earth grid of the facility. It should be understood that all of the engineering principles mentioned in this article, are wholly dependent upon having excellent soil resistivity data and valid soil models. Soil resistivity data is, and always will be, the heart of earthing science. Please refer back to the earlier section on soil resistivity.

Distribution of Current.   When lightning strikes an above earth object, such as an aerial on a LPS, the current will start to divide itself across the LPS proportionally to the impedance it encounters, on its way down to the earth. We can imagine the lightning striking the aerial and the current dividing in two as it moves from the aerial into the conductor. As the current flows down the conductor, at each conductor intersection, the current will divide again and again, until it finally reaches the earthing electrode system, where it will finally travel into the soil and dissipate. Our primary concern with the distribution of current in an earthing system is whether or not the conductors can handle the current levels, without burning open like a fuse.

Down conductors for a LPS must terminate in a connection to an earthing electrode of some kind. Most typically, it is a single standard earth rod, and on occasion three rods installed in a triangular pattern with conductors tying them together. The effectiveness of the connection to earth of each of these electrodes (or electrode systems) is measurable and is called *"resistance to earth."* The resistance to earth of any given electrode will vary, given the immediate soil conditions in which the electrode finds itself (moisture content, specific soil chemistry, proximity of non-conductive buried rocks, proximity to other conductive buried objects and/or soils, etc.), and as a result, each electrode will have a specific RTE.

As there will be multiple down conductors for any given LPS, electrodes with a lower RTE will see a proportionally larger percentage of the current. In other words, the electrodes with a better connection to the earth, will see more current than the other electrodes, "un-balancing" the LPS. The European Lightning Protection Standard EN 62305 calls for the "balancing" of these electrodes by either supplementing each electrode until they all have the same RTE, or installing a buried earth ring tying all the electrodes together. The United States has no such requirement.

When lightning enters a conductor, huge magnetic fields are formed as the energy passes through the conductor. These magnetic fields hold huge amounts of inductive energy and will induce currents into nearby metallic objects, including the same conductor (wire) it is currently travelling on. When a conductor is routed in such manner as to enable the magnetic fields from one part of the conductor to induce energy into another part of the same conductor (imagine a tight bend or circle), this is called a self-induced magnetic coupling. Self-induced magnetic couplings such as this can quickly lead to a thermal-avalanche where the two magnetic fields keep cross-coupling into each other forming a perpetually increasing energy level, thereby increasing the heat in the conductor until it melts and burns open.

All known regulatory codes regarding LPSs have detailed instructions on how to properly route conductors, so as to prevent these self-induced magnetic couplings (and thus thermal-avalanche). Not only must the straight current portion of the lightning strike be considered (impedance of the conductors, current-carrying capacity, etc.), but the magnetic fields that are formed and the subsequent current that will re-enter the system upon the collapse of the magnetic fields, must also be taken into account. Computer modelling along with good design and diligent installation techniques will prevent an overcurrent situation on any one given conductor at any point in the lightning protection system, both above and below grade.

Frequency Spectrum and Time-Domain.   It is a well-known phenomenon that lightning has both an AC component and a DC component, at the same time. In fact, lightning will propagate through a structure at many (if not all) possible frequencies. A typical lightning strike will see a range of frequencies from 0 Hz to as much as 100 MHz, and sometimes even higher. However, the distribution of these frequencies is not spread evenly and certain frequencies will be prominent. This collection of frequencies generated by a lightning strike is called the *frequency spectrum* and is primarily determined by the geometric shape of the structure struck by the lightning.

Just as the length of an antenna determines the best frequency to broadcast/receive radio signals, the same is true for lightning. The lightning will adjust its frequency based on the structure (antenna) it strikes and will resonate due to impedance imbalances between the structure and the earth. In the case of lightning, all structures are antennas. Most structures make up very complex antennas, with the buried portion adding a further complexity as it will affect the lightning strike, following many of the same principals found in the half-wave length theory for antennas. All the various variables and nuances of the calculations can become very complex, given all the different types of materials to be found on a typical structure, the erratic pathways of the LPS, and the variations found in the soil, only computer modelling can adequately calculate the expected resonant frequencies for a lightning strike on a given structure.

To calculate the frequency spectrum (or frequency response) of a lightning strike, one must first develop an accurate model of the structure with the LPS (including material types), and develop an accurate soil model for the site. Numerous frequencies must be run individually through the model, until an accurate profile can be developed. Mathematical algorithms have been developed to assist in the proper selection of the test frequencies, to reduce run times and improve the statistical accuracy of the profiles. But needless to say, many hundreds if not thousands of individual frequencies must be run through the computer simulation, to accurately determine the frequency spectrum. The final result is a graph showing the entire range of expected frequencies on the x-axis, with magnitude on the y-axis. Typically, the simulation will demonstrate that two or three frequencies will resonate through the structure during a lightning strike.

In the Time Domain, you can calculate the length of time it takes the energy from a lightning strike to clear out of your structure. The actual lightning stroke itself will start and stop in a very short time frame, typically only a few microseconds. However, as described earlier, magnetic fields will form in not only the LPS, but in the structure itself. The time it takes to generate the magnetic fields to full strength, the reflections of energy through the structure, and the time it takes to collapse the fields (steel is at least 250x more magnetic than copper and therefore holds magnetic fields for far longer) is the time duration of the lightning strike. While the actual lightning stroke may start and stop in microseconds, the time your system will be impacted by the electrical energy from the stroke, will almost certainly be many thousands of times longer as the energy generated by the strike will take time to dissipate out to the earth through the earthing system.

Conducting studies related to frequency spectrum and time domain can have substantial benefits. The primary benefits include:

1. Human Safety: Lightning strikes are different from standard electrical utility faults in not only dramatic amounts of current, but in the near limitless voltage potential and the high frequencies that will be generated. The frequency

spectrum and time domain are critical for accurately calculating the effects of a lightning strike with regards to human safety studies involving step and touch voltage hazards, EPR studies, and electromagnetic interference studies.

2. The frequency spectrum and time domain can be used in an electrical coordination and short-circuit fault study, to improve the settings of the overcurrent protection devices and reduce down-time due to unintentional power outages caused by lightning strikes. For many facilities, such as military facilities, hospitals, data centres, and power plants, a single power outage can have huge consequences and often is measured in millions of dollars in lost revenues.

3. The frequency spectrum and time domain can be used for improving surge-protection systems. While all sites will want a broad-spectrum frequency protection for unwanted surges and transients, the addition of specially tuned surge protectors designed to stop the resonant frequencies determined during the frequency spectrum study can prove especially useful for protecting vital equipment from the impact of lightning strikes.

4. In some countries, these studies are compliance requirements for international lightning protection codes (IEC 62305) without overengineering. These codes have strict rules that in many cases will simply result in over designed LPSs for many buildings. A good frequency spectrum and time domain study can prove the effectiveness of a LPS without adding huge installation costs due to over engineering. Or it will at least prove the necessity of those costs.

Additionally, once the resonant frequencies of a lightning strike have been determined, the distribution of current can be better analyzed as impedances can now be properly calculated.

Hazardous Step and Touch Voltages for Human Safety during a Lightning Strike. Human Safety is always a paramount concern, and when dealing with lightning strikes, the critical issues can compound quickly. Personnel touching a conductive object or even stepping near a lightning strike can suffer serious injury and even death. Calculating step and touch voltage hazards during a lightning strike is in principle the same as when one calculates the hazard during a line-to-earth fault. However, we must take into account the new frequencies and clearing time that will be generated during the strike, as determined by the frequency spectrum and time-domain analysis.

Step and touch voltage hazards are calculated using the strike amperage at the point of contact, the frequency, the X/R Ratio, the clearing time (i.e., time domain), and the specific soil resistivity conditions. Only a computer simulation can accurately model the frequency spectrum and time domain of the lightning strike to determine the touch and step voltages that will be experienced during a strike. Once the safety parameters (crushed rock, shoes, etc.) have been applied to the computer model, the overvoltage hazards will appear and standard mitigation techniques for reducing these hazardous voltages can be applied, thus making the site safe.

Electromagnetic Interference and Earth Potential Rise during a Lightning Strike. Another key factor during a lightning strike is electromagnetic interference. As mentioned earlier in this article, large magnetic fields will form in both the LPS and in the buried earthing/earthing system. Additionally, the EPR will cause scalar voltages to form across the surface of the earth with the potential decreasing with distance. These large magnetic fields and EPR effects will transfer voltages and currents (by inductive, capacitive and through earth coupling) into nearby pipelines, railways, communication lines, homes, industrial facilities, farms, and other such utilities, whether buried or aboveground.

The current and voltages that are induced into the nearby utilities can cause great damage. One of the primary concerns is on the induced voltages/currents that can form on the data and/or shield lines of communications cables causing damage as these transient currents will flow through sensitive electronic equipment on their way to earth. The neutral wires from the utility power company can carry stray currents formed from the electromagnetic fields up and into homes and industrial parks. Nearby gas pipelines can have their protective coatings compromised by the stress-voltages caused by the difference in potential between the surrounding earth and the pipe.

Again using the highest magnitude frequencies of the lightning strike, as determined by the frequency spectrum analysis, the electromagnetic fields and the EPR can be accurately calculated and the impact on the surrounding infrastructure can be properly analyzed and potential problems mitigated.

Summation of Lightning Effects on Earthing Systems. When calculating the effects of a standard electrical utility fault on an earthing system, we know that the utility fault will have clamped voltages at a very specific frequency (50/60 Hz). With a lightning strike, the potential voltages are virtually unlimited and will have multiple frequencies based upon the geometric shape of the structure that is struck. The hazards presented by an electrical fault at 50/60 Hz are very different from the hazards presented by an electrical fault at 50 kHz or even 60 MHz. These order-of-magnitude increases in frequency present unique engineering challenges and is why lightning strikes are so dangerous. The bottom line is that EPR and step and touch voltage hazard studies conducted using power company (utility) fault data are simply not applicable for a lightning strike.

So, in summation, all of the factors presented above need to be taken into account when building an effective LPS, be it a utility substation, telecommunication site, or simply just a personal home.

STANDARDS FOR LIGHTNING PROTECTION SYSTEMS

There are significant differences between the U.S. Standards and the International standards for lightning protection. The international standards such as the *IEC 62305-3* and its British counterpart the *BS EN 62305* have many additional requirements above and beyond those found in the U.S. standard *NFPA 780*. The differences between these standards are worth noting, for those who are interested in providing lightning protection for their structures.

Lightning Protection Systems. Lightning protection systems are not only one of the most important (and expensive) infrastructure components of a building, but is also one of the least understood. Poorly designed LPSs can add unnecessary expense, add legal liabilities to your construction project, and may not provide the protection you need. A good lighting protection system is engineered and designed specifically for your structure and the needs of your facility. In the United States, most industry and the government facilities are protected by *NFPA 780 Standard for the Installation of Lightning Protection Systems*. This standard is issued by the *National Fire Protection Association* (NFPA), the same group that writes the *National Electrical Code*, and provides a guideline for installing a one-size-fits-all LPS. In Britain, LPSs are build to be in compliance with BS EN 62305.

U.S. Standard NFPA 780.   *NFPA 780* provides guidelines for how often to place air terminals, spacing for cross and down conductors, earth rod and loop requirements, surge-protection requirements, and how to install protection for trees, towers, and so on. The standard, however, has two primary shortcomings, in that it does not analyze the installed systems ability to handle an actual lightning strike, nor does it take into consideration what the system is protecting. In other words, *NFPA 780* has the same requirements for a garage as it does for a billion-dollar computer farm. These shortcomings along with virtually no legal and/or insurance industry requirements for lightning protection, has led many facilities managers to simply take their chances and forgo protecting their buildings.

International Standard IEC 62305.   We can learn a lot about lightning protection by looking at the requirements of *IEC 62305* which is significantly more demanding than the U.S. NFPA 780 standard. Here are the basic requirements of *IEC 62305*:

1. It requires a risk factor assessment that determines the level of required lightning protection. This risk assessment is quite complex and software is almost always needed to make a proper assessment. The calculation includes human life, public services, cultural heritage, economic risk, and occupancy issues. This risk assessment is both good and bad in that a garage will have fewer requirements than found in NFPA 780, but billion-dollar computer farms will have greater requirements.

2. The *IEC 62305* standard requires an actual assessment of the LPS to insure that it is capable of handling a lightning strike. The lightning strike calculations are far more significant for both the time domain parameter and the actual strike amperages (100–200 kA), than the U.S. industry standard (often only 15 kA). Calculations that are often required include:

   a. Expected amount of current to be carried on individual conductors in DC amps, to ensure that current carrying capacity is not exceeded

   b. Rolling-ball theory of lightning protection tested, against 3D computer models of the structure and surrounding area

   c. Spark gap and arc-flash calculations from the LPS to adjacent conductive utilities

   d. Time-domain of the lightning strike on the specific structure. This is critical to understanding the amperage carrying capacity of the conductors. Without an actual calculation, the *IEC 62305* default time-domain could force the unnecessary installation of additional conductors

   e. Frequency spectrum of the lightning strike on the specific structure. This data is needed for both surge-protection and for timing of circuit breakers to prevent power outages

3. In general, *IEC 62305* has physical construction and installation requirements that are far greater than *NFPA 780*.

   a. Down conductors vary from 10-m to 20-m spacings. The *NFPA 780* use one-size-fits-all 30-m spacing.

b. The zone of protection or rolling-ball theory in the IEC standard varies the angle required based on the risk assessment, which can impact placement of certain types of aerials from 20 to 60 m heights.

c. Concrete columns that are used for down conductors must be tested a 0.2 Ω or less continuity, and rebar must be welded with 20x diameter overlaps. These must be bonded to the floor slab.

d. Earth rings are required for all non-conductive buildings, buildings housing electronic systems, and certain risk factors. Individual rod installations (without earth rings) must be tested, so that each electrode is at the same RTE.

e. Spark gaps between lightning conductors and other metallic objects must be considered.

f. Incoming utility services (such as overhead power lines) and adjoining public spaces may also be required to have protection systems installed, based on the risk assessment.

g. Both internal and external lightning surge protection systems are mandatory.

4. The *IEC 62305* has stringent requirements for annual testing and inspection of the LPSs. This of course goes along with mandatory maintenance requirements.

5. The *IEC 62305* standard gives you three choices when it comes to selecting an interception model, we recommend you use the rolling sphere model (RSM) or sometimes known as the electro-geometric model (EGM). The protection angle method (PAM) and the mesh method (MM) should not be used. These are legacy methods that have been left in the standard due to historical reasons; you will be far better served by using the RSM.

Lighting Protection System Recommendation. It is recommended to look at both the *NFPA 780* and the *IEC 62305* for guidance. Using a combination of the two, including a reasonable risk assessment and computer modelling, a good engineering firm will be able to maximize human safety and cost effectiveness when designing your LPS. Advanced computer-modelling techniques which include the rolling sphere method, frequency spectrum and time-domain analysis, and step and touch voltage Analysis will ensure that your facilities needs are met without over or under engineering your lighting protection systems.

Chapter Ten

ECONOMIC AND LEGAL ANALYSIS

Note: This chapter, written by Christopher Clemmens, is based on a U.S. legal analysis; however, the principles largely apply to British law as well.

ECONOMIC CONSIDERATIONS

Economics studies the allocation of scarce resources and the trade-offs that entails. If a nearly infinite amount of resources were applied to grounding and otherwise idiot-proofing high-voltage electrical equipment, it is theoretically possible to reduce the number of deaths due to electrocution from high-voltage electrical equipment to zero. But by so doing, resources would be starved from other areas, such as health care or policing, and the overall number of societal deaths would increase.

On the level of a company, if a company does not invest enough in grounding then it could open itself up to damaging litigation and if it invests too much it will no longer be competitive with other firms. So, a rational choice has to be made that balances potential harm to the firm from an accident against the "harm" of investing in protective measures that do not directly generate revenues. In addition, a company has to factor in how much it is willing to spend to maintain its reputation as a trusted service provider and a member of its community. For example, an electrician should not just think about how much he would be out of pocket if faulty grounding work caused a fire, but he should also consider how that fire might inconvenience others and cause damage to his own reputation. And beyond purely economic considerations, no ever wants to carry the burden of causing harm to others because of substandard or otherwise shoddy grounding work.

Calculating rational investment levels based on potential damage to equipment is the easiest part of the calculation. The traditional method is to multiply the cost of equipment by the likelihood of failure. For example, if one has a $100,000 piece of equipment and a grounding failure could cause a 1% chance of failure per year, then the baseline investment in grounding would be $1000 yearly. This assumes that the company has the financial wherewithal to withstand the financial loss and/or that the loss of equipment would not cause a failure in customer service.

If the financial loss to the firm would be catastrophic, then a larger investment in grounding and/or insurance would be called for. But it will be difficult to collect an insurance claim if inadequate investment in grounding leads to negligence on the part of the insured. Almost all insurance policies only cover accidents and have clauses about not covering intentional actions and negligence. This makes sense from the perspective of the insurer because it wants to avoid moral hazard that is, taking unnecessary risks because the insured party knows that the insurer will pay. Even if one can get an insurance company to pay for damages, a bad record will lead insurance companies to raise premiums to crippling levels.

Aside from catastrophic grounding damage, there are also slower ways through which equipment is damaged, such as corrosion. A key tool in cutting corrosion costs is grounding. In 2002, the U.S. Federal Highway Administration (FHWA) released the first, and so far only, congressionally mandated study on the direct costs associated with metallic corrosion in the economy as a whole. This thorough study took two years and covered nearly every sector of the U.S. economy. Results of the study show that the total annual estimated direct cost of corrosion in the United States is a staggering $276 billion—approximately 3.1% of the nation's gross domestic product (GDP) at the time of the study. The study estimated that if indirect costs such as outages, delays, failures, and litigation were factored in, total costs would be about 6% of GDP. In today's dollars direct cost would be about $1500 for every American, and if one included indirect costs then per capita costs would approach $3000. The study found that 25% to 30% of these costs could be avoided if optimum corrosion prevention methods are used. Further, it showed that corrosion-control methodologies are making slow but steady progress. A Department of Defense study found that corrosion costs over $2.5 billion and over 20% of total maintenance costs.[1] It also significantly hurts military readiness because it sidelines crucial equipment for maintenance and repair.

Of course, the real concern with grounding safety is not loss of equipment but loss of life. Unfortunately, over 400 Americans die from electrocution in the workplace each year and another 500 die every year in house fires caused by electrical malfunctions. There are different methods for calculating the value of a life. Insurance claims and wrongful-death lawsuits look at present value of lifetime earnings, potential contribution to society, and emotional distress caused to loved ones. To this, courts will often add punitive damages if they find that the defendant is negligent.

Government agencies also calculate the value of a life when they are setting safety and other regulatory standards. Then, they calculate whether to mandate a safety feature using a similar calculation that we used above with the cost of equipment. If the value of a life is deemed to be $7 million and a highway guard rail would have a 1% chance of saving a life, if the cost of the guard rail was under $70,000, its use would be mandated. If it is over $70,000, its use would not be mandated. This calculation might seem quite grim, but we have to remember that if we mandated a crippling amount of investment in safety, then we would have no money for other areas of the economy. Determining the value of a life is an inexact science. It varies from agency to agency and from one period of time to another. It generally increases, but it can also decrease as when the U.S. Environmental Protection Agency (EPA)

---

[1]Under Secretary of Defense for Acquisition, Technology, and Logistics, *DoD Annual Cost of Corrosion*, July 2009.

dropped the value of a life from $7.8 million in 2003 to $6.9 million in 2008. In the same period that the EPA cut its values, the Department of Transportation raised its values twice. The value of a life also varies by country. According to a 2012 Rosgosstrakh Strategic Research Centre survey, the median value of a life in Russia is only $44,700. One can only hope that responsible parties are using their consciences, and not just cold financial calculations, when making grounding decisions in Russia.

The indirect costs for a firm or tradesman of causing injury or death are extensive. Reputational harm might result in having no more customers. A firm involved in an accident may be seen as a dangerous or otherwise undesirable employer. Management may become so distracted by dealing with the fallout that it falls into a reactive management mode rather than focusing on building its business. Worse yet, firms might be dragged down by heavy-handed regulatory agencies and crippling lawsuits.

LEGAL CONSIDERATIONS

Of course it is beyond the scope of this book to offer legal advice nor are the authors qualified to do so. That said, electrical grounding problems can damage equipment and even take lives. This can lead to damaging liabilities that can ruin the finances of tradesmen and companies. Also, when discussing legal issues, one must be aware that regulations vary from country to country and even from jurisdiction to jurisdiction within a country. For example, in some U.S. states it is possible for tradesmen to discharge liabilities caused by negligence through the process of bankruptcy, while in other states it is not.

Almost all legal actions taken against electrical companies occur in civil, as opposed to criminal, courts. A key distinction is that in a criminal case the plaintiffs must prove their case "beyond a reasonable doubt," which means that the judge or jury is almost certain of the defendant's guilt. In civil courts, the plaintiffs must only prove their case by a "preponderance of evidence," which means that the defendant is more than 50% likely to be liable for whatever damages the plaintiff is claiming. Under normal circumstances, this means that the defendant caused damages and failed to exercise the care that a "reasonable person" would use in similar circumstances. In other words, there were damages and the damages were likely caused by the defendant's negligence.

> "The core idea of negligence is that people should exercise reasonable care when they act by taking account of the potential harm that they might foreseeably cause harm to other people."[2]

Although the "reasonable person" standard is the normal benchmark for determining negligence in civil cases, there are circumstances where the defendant is held to a higher standard. For example, a certified electrician would be held to a much higher standard of due care than a layman. Another relevant example would be that a company that deals in ultrahazardous materials, which in many circumstances electricity could be considered, is held to a much stricter standard than a company dealing in less inherently dangerous goods.

---

[2]Feinman, Jay (2010). *Law 101*. New York: Oxford University Press.

Issues of electrical grounding span many industries, and there are thus different standards of legally required due care. Most of the early legal precedents for electricity and the law were set by cases involving electric transmission companies. Electric transmission companies are held to the "standard of foreseeability." There are two elements to this standard. The first is that defendants are held liable for "anticipating all proof is heavily stacked against a company eventualities." This is in contrast to the normal common law standard, which holds plaintiffs only to the standard of taking precautions that a "reasonable person" would take in similar circumstances and places the burden of proof on the plaintiff to prove that the defendant did not fulfil a burden of due care. The standard of foreseeability shifts the burden of proof from the plaintiff to the defendant, so the defendant must prove that there was no way that the mishap could have been foreseen. Given the long and deep case law associated with the use of electricity and grounding, the burden of proof is heavily stacked against a company arguing that a situation that caused damages was not foreseeable. The Occupational Safety and Health Administration's general industry electrical safety standards published in *Title 29 Code of Federal Regulations (CFR)*, Title 29, Part 1910.302 through 1910.308—Design Safety Standards for Electrical Systems—and Part 1910.331 through 1910.335—Electrical Safety-Related Work Practices Standards, leave few conceivable areas where those operating in electrical industries are not responsible.

The second element of the standard of foreseeability is that the defendant must have taken every reasonable action to prevent the damage. A defence under this standard might be force majeure. Force majeure is Latin for "superior force" but it is often translated as "an act of God". In other words, a force that was completely outside of the defendant's hands. For example, if a plane crashed into an electrical line and the electric company had taken every precaution to make sure that electricity is cut off to severed wires, but the live wire still caused a fire or injury, the electric company might escape having to pay damages. Force majeure defences are anything but an easy out for defendants. Courts scrutinize force majeure claims intensely and do not allow them unless the circumstances were truly unforeseeable and unpreventable. For example, damages caused by a tornado in Oklahoma or an earthquake in California would not qualify as force majeure because these are common enough occurrences that they should be anticipated.

A related standard applied to companies involved in electrical industries goes back to at least 1912:[3] *res ipsa loquitur* (Latin for "the thing speaks for itself"). What this standard effectively does is to shift the burden of proof from the plaintiff to the defendant. Or as stated in *Tassin v. Louisiana Power & Light Co.* "the burden shifts to the defendant to show that the accident was caused by something for which it is not responsible".[4] Under this standard, the plaintiff does not need to prove that the defendant breached a duty or acted unreasonably. The plaintiff only needs to prove that it suffered harm and that the harm was caused by the defendant. An interesting example of this principle is *Hoffmann v. Wisconsin Electric Power Company*. The plaintiffs got $1.2 million because they proved that poor grounding work on the part of Wisconsin Electrical Power Company caused a low level of stray voltage [a ground potential rise (GPR) event] over a long period of

[3]*San Juan Light & Transit Co. v. Requena*, 224 U.S. 89 (1912).
[4]*Tassin v. Louisiana Power & Light Co.*, 191 So. 2d 338, 341 (La. App. 3d Cir. 1966).

time on their dairy farm. They claimed that this harmed their dairy farm production. Although Wisconsin Electric Power Company was not in violation of state laws, it had to pay damages nonetheless.

The highest legal standard of liability, strict liability, applies in the cases of stray voltage accidentally discharged into the ground. Strict liability applies in situations that the law deems inherently dangerous. Strict liability is best understood through a classic legal hypothetical. Wild animals, such as lions, are legally considered inherently dangerous. Therefore, an owner of a lion would be liable for the damage that an escaped lion caused regardless of how sturdy a cage the lion was kept in. Under strict liability the plaintiff need not prove that the defendant was negligent. The plaintiff only needs to prove that damage occurred and that the damage was caused by the defendant. For example, if a child were harmed due to a grounding problem, the defendant would be prima facia guilty. This standard of law forces defendants to take every possible precaution as opposed to the standard burden of taking "reasonable" precaution. Strict liability applies to cases where stray electrical voltage is discharged into the ground. Therefore, the only legal defence against damages caused by grounding problems is to ensure the absolutely highest quality of grounding that prevents accidents.

In some situations, contributory negligence on the part of the defendant can be used as a defence. Contributory negligence is when a plaintiff, as a result of its own negligence, partially causes its damages. To the extent that a plaintiff contributes to its own damages, compensatory damages, but not punitive damages, will be reduced. A classic example of this would be a woman who is drying her hair while in the bath and is electrocuted when she drops the hair dryer into the water. Given that builders now have to install ground fault circuit interrupters in bathrooms, courts would most likely find that fault to be partially, but not wholly, that of the bather. Contributory negligence will not insure that courts will not impose severe losses upon a defendant. In the famous McDonald's coffee case, *Liebeck v. McDonald's*, the plaintiff was found to be 20% responsible for the damages and this reduced the compensatory damages from $200,000 to $160,000. But that did not stop the jury from awarding her an additional $2,700,000 in punitive damages. Although, McDonald's ended up paying a lesser, undisclosed amount after appeals, the settlement, litigation costs, and reputational damage were significant even though Ms. Liebeck was found to have contributed to her own damages.

Companies that employ trained professionals are not free from the concerns of legal liability. Professionals working around electrical equipment are expected to practice due care[5]. As mentioned above, over 400 Americans die from electrocution in the workplace each year and another 500 die every year in house fires caused by electrical malfunctions. But even when an employee is at fault, litigation is expensive, the publicity of a court case can cause significant reputational damage, and juries and judges cannot be relied upon to find for even a reasonably cautious defendant.

Of course not all electrical equipment can be kept in a controlled environment that can only be accessed by professionals. Extreme levels of caution must be taken when the general public might have access to equipment. This is especially true when children are involved because children are held to a different standard of

---

[5]*Butler v. City of Peru*, 733 N.E.2d 912 (Ind. 2000).

reasonableness when it comes to assigning contributory negligence.[6] A company might not be found liable if an adult breaches a fence and trespasses onto a substation, but in Dungee v. Virginia Electric & Power Company the defendant had to pay $20 million in damages when a 10-year-old boy received severe burns when he snuck into a substation to retrieve a ball. And aside from the ethical reasons for protecting children, damages are usually greater in cases involving children because a key variable in value-of-life calculations is the number of years of life lost, which is clearly higher with children. Further, juries can be swayed by emotion and they are likely to impose much higher punitive damages when children are involved.

In an action for injury from electricity, the question of whether the injured person was guilty of contributory negligence is usually left to the consideration of the jury. A person's intentional conduct in exposing himself or herself to electricity can supersede any alleged negligence or wantonness of the power company, thus precluding liability. The law imposes on a person sui juris, the obligation to use ordinary care for his or her own protection, and the degree of such care should be commensurate with the danger to be avoided. For example, a reasonable person knows not to stick a fork into an electrical outlet. Although the perpetrator of this act may win a Darwin Award, it is unlikely that his surviving family members would receive a settlement.

In conclusion, given the engineering complexities, the high cost of copper, and the labour expense of installing extensive below-grade electrical grounding systems, careful considerations must be taken into account when making decisions regarding those systems related to human safety and effective equipment operations, such as earthing and grounding. Some of the issues the company executive must consider are:

- Legal responsibilities of the company
  - The safety of employees and contractors
  - The safety of the public
- The cost of life
- Legal liabilities of the company
  - Negligence: anticipating all eventualities
  - Negligence: standard of foreseeability
  - Strict liability (when dealing with inherently dangerous substances such as electricity)
- Effective equipment operation
  - Cost of equipment failure
  - Cost of increased equipment maintenance
  - Data and communication losses
  - Corrosion

For most executives, the decision to hire third-party engineering firms that specialize in grounding and earthing, to design and develop cost-effective and efficient grounding systems is fundamental in mitigating the risks a company and its employees face.

---

[6]Toropdar v D (2009) EWHC 2997.

Chapter Eleven

MANAGING A GROUNDING AND EARTHING PROJECT

Note: It is seldom that engineering issues make the news and even rarer when the media covers earthing issues. The following story attracted the attention of the major news networks in the United States, bringing a new level of pressure to electrical manager Joseph A. Anderson. TV trucks tend to do that. This chapter, written by Mr. Anderson himself, may provide you with some valuable insight on managing your next earthing project.

For an electrical manager, overseeing a grounding and earthing project can require overcoming many underlying obstacles. I manage the electrical department of a 900-acre campus comprising nearly 400 buildings ranging from residential to industrial status. Many people refer to the campus as a small city because it has been designated its own postal zip code. Although it is not technically a city it does have an economic impact on the community that is as impressive as the campus itself. As you may imagine, keeping an electrical infrastructure of this magnitude operating safely and efficiently can be a constant undertaking. As a master electrician with over 11 years in the field, I've found that grounding and earthing projects are among the most trying. Whether you start grounding and earthing projects to simply make improvements or you start one because of problems that arise, there are many considerations that must be taken into account. Although it is impossible to cover all of the circumstances encountered with these projects, it will help to have an idea of points that must be thought out. I will start by sharing an experience of a challenging grounding and earthing project.

THE STORY

I was called to the campus swimming pool at the facility I had worked at for many years. General maintenance workers had discovered a problem with the pool pump motor. While troubleshooting the cause of failure I discovered a transient voltage and current that was actually travelling through the pool water. This was especially startling to me because people were swimming in water containing measurable voltage. If you're reading this book, you probably know that water and electricity do not mix.

I had no choice but to calmly inform a lifeguard that the pool needed to be evacuated immediately and closed until further notice. This problem was discovered in the middle of July at the height of swimming season. The pressure to isolate the problem promptly and get the pool up and running was unbelievable. However, I knew the depth of the problems that made this condition possible were much deeper than anyone could anticipate. The pool was constructed in 1964 prior to the area's adoption of the 1962 National Electric Code. Therefore, it had been constructed with a poor grounding system, and was completely lacking the now required equipotential plane for swimming pools (NEC 680.26). Even though the problem that created this condition could be corrected, there was no system in place to protect swimmers in the event of reoccurrence.

At this point I was faced with some troubling decisions. I was stuck between the needs of the organization, the frustration of management, as well as liability that the organization and I could face. Negligence should always be considered by organizations and professionals. In this case, it was my opinion that to simply repair the current problem and reopen the pool, it would be just that negligent. Any number of unforeseen electrical failures could present hazards that would put the public at risk. Both ethically and morally, I was left with no choice.

I informed my manager in writing of the possible hazards, financial impacts, and durational demands of the renovation needed. At this point, frustration within the organization increased. Everyone, from managers to the president, questioned how the pool could be open since 1964 and now was suddenly unsafe. I felt pressure to make recommendations for repair and to move forward. However, I knew that it was out of scope for a master electrician to develop a plan of this magnitude. I made a strong recommendation to involve an electrical engineer.

I was starting to feel like I was on the chopping block, and to make matters worse, I feared that once an electrical engineer was involved the pool would be deemed unrepairable. It was very important to partner with an electrical engineer who could fully grasp the problem at hand as well as fully understand any and all possible solutions. After all, the ramifications of not aligning with a true professional left the possibility of a permanent closing of the pool. I sought out an expert in this specific area of practice who I believed could make the proper recommendations.

After expanding my search out of state, I found an electrical engineer who had an optimistic outlook on repairing the pool. Moving beyond local engineers may have had a financial impact to the front end of the project, but I believe that it had positive impacts to the end result. Excited and relieved about his views, I promptly scheduled a consultation. Expressing the urgency of our needs, he agreed to fly out the following week.

After finding an appropriate electrical engineer, I immediately pulled a state electrical permit and started on improving circuitry grounding throughout the building. I really wanted to improve the building's electrical system as much as possible before the engineer's assessment. All of the circuitry within the pool building relied on conduits for grounding. Although this is accepted by code, it is not a recommended practice in the industry. This is what caused the initial problem with the motor at the pool. The faulty motor had lost its ground thus sending fault currents through the water. So that is where I started. I brought in all qualified personnel from within the organization as well a local electrical contractor. Ground wires were pulled through all conduits providing a proper, solid return for all fault currents. Also, Edison-type circuits (shared neutral circuits) were eliminated by adding neutral wires for each circuit. Then GFI protection was added in all required areas of the building (nearly 100%). This was roughly 60% complete by the time the electrical engineer arrived.

Upon arrival of the engineer, an in-depth analysis of the safety of the swimming pool began. The analysis included root cause determination, Wenner four-point soil resistivity testing, continuity testing, ground system resistance testing, a point-to-point (two-point) resistance test on all metal objects around the pool, and a ground potential rise (GPR) study. After completion of the analysis, the dangers of the pool were evident. It was much easier to explain to management once the test results were in front of us on paper. Frustrations and tensions began to fade into thankfulness. We were all thankful that no one was ever hurt. As a result, I gained support from the management and I was able to move forward with the electrical engineer. A full report was provided as well as a final recommended grounding design for improving the pool.

Since it was highly unlikely that the 50-year-old pool could survive a complete retrofit to meet current standards, a "back bone" grounding system was recommended. The proposed "back bone" grounding system would bond to the steel rebar in the concrete foundation in seven places including the corners of the pool. This provided a new grounding electrode system that would ensure the system was tied to ground. It would also bond to metal piping, structural steel, water pipe, and pump motors to form a new equipotential grounding system. All of this was to be bonded to nine ground rods in triad formation at key locations outside the building. The plan was taken directly to the state electrical inspector for preapproval.

After preapproval, the hard work began. Luckily, the pool was constructed with a crawl space below the deck for maintenance. First, I prepared personnel to expose rebar at seven places under the pool. They also needed to expose connection points at all pool ladders and railings. They used a metal detector and jackhammer to accomplish this. Second, I prepared an accurate material list and ordered material. Since cad welding the rebar would be difficult, I also needed to purchase some specialty tools to terminate compression fittings. Third, I scheduled a welding contractor to weld proper-grade rebar (W-grade) at the seven exposed points for connection. Next, I scheduled an excavation crew to prepare trenches for connection to the ground triads. Last, I coordinated and assisted in completion of all the remaining work. This included completion of improvements to the existing circuitry, running ground wire full circle around the entire building (above-grade equipotential ground ring), mounting ground bars for terminations at key locations throughout the run, connection to all metal parts near the pool, connection to the exposed rebar and all pool-related equipment, driving ground rods for grounding triads, and connection to the triads, including test wells.

Nearly one month after the closing of the pool, the work was complete, we were ready to test the system, and the electrical engineer was on site to verify the work.

Upon testing, it was confirmed that voltage gradients throughout the building had been reduced to acceptable levels. Also, all of the stray voltages and currents were gone. All of the repairs and improvements had worked. The state electrical inspector was scheduled for a final inspection and the electrical engineer and I met him to perform the testing again. The inspector was pleased with the test results as well as with all of the work completed. The pool was reopened the next day and was safer than it had been in its 50-year history. Everyone was pleased and I actually received the employee of the month award after all was said and done.

LESSONS LEARNED

I believe that this story illustrates some points to consider when managing grounding and earthing projects. I will continue by expanding on particular points that were considered in this project. First, you must consider what you, your manager, and electricians in the field need to know. Second, you must consider the overall goals and demands of the project. Last, you must consider limitations and benefits of products and materials used on the project.

First, you must consider what you, your manager, and electricians in the field need to know. This will get the ball rolling on starting the project. As an electrical manager you will need to be proactive about:

- The safety of the public and personnel
- Protecting yourself and the organization you represent
- Identifying hazards
- Assessing risks
- Communication
- Properly informing management of hazards
- Gaining the support of management

Once you have identified a hazard it is your due diligence to communicate it to management. I suggest clear, concise communication in both written and verbal forms. Verbally expressing the dangers will give you a chance to gain support in mitigating the hazard. Written communication will give you proper documentation that management was made aware of the existing conditions. If you do gain support, you must communicate the importance of the project to electricians in the field. They must be well informed of the problem at hand, expected outcomes, and all of the steps in between.

Second, you must consider the overall goals and demands of the project. This includes both financial demands and time constraints. This was especially challenging with the pool project because the full depth of the project was unknown before we were provided with the electrical engineer's grounding design. Serious estimating consideration was crucial to properly inform management of the initial project demands. Extensive time was spent on research to determine:

- Proper material
- Material cost
- Labour demands
- Duration of time the project will consume
- Engineering needs

- Excavation needs
- Contracting needs
- Tools needed(e.g., renting versus purchasing compression tools)
- Electrical inspections and permit costs
- Necessary safety equipment and procedures

Also, you must set clear goals regarding project expectations as well as a reasonable timeline to accomplish these goals. This was especially important with this project because every day that the pool was closed the financial commitment increased. For example, the organization paid the township daily for usage of their pool by our members and guests. Not to mention the reinspection fees that would be assessed if we were not on schedule with the state electrical inspector.

Last, you must consider the limitations and benefits of products and materials. For example, I had to weigh the benefits of cad welding versus the benefits of irreversible compression fittings for terminations at the pool. Since copper and steel have different melting points, cad welding can be difficult when working with rebar. So, compression fittings were used on this particular project. Also, I found that particular grades of rebar are suitable for electrical connection and some are not. In the case of the pool, I had to bring in a qualified welder to add small lengths of W-grade rebar to the existing rebar to make a suitable point for connection. Then the use of an oxide-inhibiting compound (penetrox) was considered. An oxide-inhibiting compound is commonly used on aluminium terminations. The wire used at the pool was all copper, but given the corrosive environment and the importance of solid connections, an oxide-inhibiting compound was used that was designed especially for copper. Potting compound was also an important consideration. Termination of electrical connections near pool water was one thing the electrical inspector was particularly concerned with. Potting compounds help to insulate connections and protect them against high humidity and corrosive environments. The ground rods used in the pool project were also considered. Ground rods come in various diameters and lengths. Approximately 10 ft 5/8 in. ground rods were used in triad formation. This required spacing them 20 ft apart from each other as to not overlap their sphere of influence. The sphere of influence of a ground rod is equal to its length. All of these are important considerations when managing a grounding and earthing project.

In conclusion, overseeing a grounding and earthing project can require overcoming many underlying obstacles. Although it is impossible to cover all of the circumstances encountered with these projects, it will help to have an idea of points that must be thought out. First, you must consider what you need to know, what your manager needs to know, and what electricians in the field need to know. Second, you must consider the overall goals and demands of the project. Last, you must consider limitations and benefits of products and materials used on the project.

REFERENCES

1. Beaty, H. Wayne. *McGraw-Hill's Handbook of Electric Power Calculations*, 3d ed. New York: McGraw-Hill. https://www.mhprofessional.com.
2. Fink, Donald G., and H. Wayne Beaty. 2013. *Standard Handbook for Electrical Engineers*, 16th ed. New York: McGraw-Hill. https://www.mhprofessional.com/.
3. Stockin, David R. 2014. *McGraw-Hill's National Electrical Code 2014 Grounding & Earthing Handbook*. New York: McGraw-Hill. https://www.mhprofessional.com/.
4. ANSI/IEEE Standard 142-1982, IEEE Recommended Practice for Grounding of Industrial and Commercial Power Systems (Green Book), ANSI/IEEE, 1982. http://ieeexplore.ieee.org.
5. IEEE Standard 367-1996, IEEE Recommended Practice for Determining the Electric Power Station Ground Potential Rise and Induced Voltage from a Power Fault, IEEE, 1996. http://ieeexplore.ieee.org.
6. NFPA 70, National Electrical Code, National Fire Protection Association, Quincy, MA, 2010. http://www.nfpa.org/.
7. National Fire Protection Association. 2014. *NFPA 70: National Electrical Code Handbook*, 13th ed. Quincy, MA: NFPA. http://www.nfpa.org/.
8. NFPA 780-2011, Standard for the Installation of Lightning Protection Systems, Quincy, MA, 2011. http://www.nfpa.org/.
9. OSHA General Industry Standards, Subpart S, Electrical. https://www.osha.gov.
10. Safe Engineering Services and Technologies, home page [Internet grounding reference]. www.sestech.com.
11. ANSI/IEEE Standard 81-1983, IEEE Guide for Measuring Earth Resistivity, Ground Impedance, and Earth Surface Potentials of a Ground System, ANSI/IEEE, 1983. http://ieeexplore.ieee.org.
12. ANSI/IEEE Standard 81-2012, IEEE Guide for Measuring Earth Resistivity, Ground Impedance, and Earth Surface Potentials of a Ground System, ANSI/IEEE, 2012. http://ieeexplore.ieee.org.
13. ANSI/IEEE Standard 80-2000, IEEE Guide for Safety in AC Substation Grounding, ANSI/IEEE, 2000. http://ieeexplore.ieee.org/.
14. ANSI/IEEE Standard 80-2013, IEEE Guide for Safety in AC Substation Grounding, ANSI/IEEE, 2013. http://ieeexplore.ieee.org/.
15. Lee, Bok-Hee, Jeong-Hyeon Joe, and Jong-Hyuk Choi. 2009. "Simulations of Frequency-Dependent Impedance of Ground Rods Considering Multi-Layered Soil Structures." *Journal of Electrical Engineering & Technology*, Vol. 4, No. 4, pp. 531–537. http://home.jeet.or.kr/index.asp.

16. Martins, António, Sílvio Mariano, and Maria do Rosário Calado. 2012. "The IEEE Model for a Ground Rod in a Two Layer Soil—A FEM Approach." *Finite Element Analysis—New Trends and Developments.* InTech. October 10. http://dx.doi.org/10.5772/48252.
17. Dwight, H. B. 1936. "Calculation of Resistances to Ground," *AIEE Transactions,* Vol. 55, pp. 1319–1328, December.
18. Wenner, F. 1916. "A Method of Measuring Resistivity, National Bureau of Standards." *Scientific Paper,* Vol. 12, No. S—258, pp. 469.
19. Canadian Electrical Code 2012. 2012. *Safety Standard for Electrical Installations,* 22d ed. C22.1-12, Canadian Standards Association.
20. Wightman, W. E., F. Jalinoos, P. Sirles, and K. Hanna. 2003. *Application of Geophysical Methods to Highway Related Problems.* Lakewood, CO: Federal Highway Administration, Central Federal Lands Highway Division. Publication No. FHWA-IF-04-021, September. http://www.cflhd.gov/resources/geotechnical/documents/geotechPdf.pdf.
21. Wikipedia. "Telegrapher's equations." http://en.wikipedia.org/wiki/Telegrapher%27s_equations.
22. http://www.express.co.uk/news/nature/490411/Lightning-DOES-strike-twice-and-other-shocking-facts.
23. BS 7354:1990, Code of practice for design of high-voltage open-terminal stations.
24. BS EN 61936-1:2010+A1:2014, Power installations exceeding 1 kV a.c. Common rules. http://shop.bsigroup.com/ProductDetail/?pid=000000000000413162.
25. BS EN 50522:2010, Earthing of power installations exceeding 1 kV a.c. http://shop.bsigroup.com/ProductDetail/?pid=000000000030270717.
26. BS IEC 60050-195:1998, International electrotechnical vocabulary. Earthing and protection against electric shock. http://shop.bsigroup.com/ProductDetail/?pid=000000000001576006.
27. International Electrotechnical Vocabulary, IEV 195.05.12. Area: 195: Earthing and protection against electric shock/Voltages and Currents, Step Voltage, http://www.electropedia.org/iev/iev.nsf/display?openform&iev ref=195-05-12.
28. International Electrotechnical Vocabulary, IEV 195.05.11. Area: 195: Earthing and protection against electric shock/Voltages and Currents, Touch Voltage, http://www.electropedia.org/iev/iev.nsf/display?openform&iev ref=195-05-11.
29. ENA TS 41-24, Guidelines for the design, installation, testing & maintenance of main, earthing systems in substations, http://infostore.saiglobal.com/EMEA/Details.aspx?ProductID=229983.
30. IEC TS 60479-1:2005+AMD1:2016 CSV Consolidated version, Effects of current on human beings and livestock, Part 1: General aspects.

INDEX

9 781259 641275